Zin Mar Than
Socio-Economic Development of
Indawgyi Lake, Myanmar

T0139748

URBAN AND REGIONAL DEVELOPMENT IN MYANMAR

Edited by Frauke Kraas / Germany, Aung Kyaw / Myanmar,

Martin Coy / Austria, Shigeko Haruyama / Japan, Boon-Thong Lee / Malaysia,

Nay Win Oo / Myanmar and Zin Nwe Myint / Myanmar

Volume 1

Zin Mar Than

Socio-Economic Development of Indawgyi Lake, Myanmar

Franz Steiner Verlag

This publication was printed with financial support of the Deutscher Akademischer Austauschdienst (DAAD, German Academic Exchange Service)

Cover illustration:
Fisherman on the Indawgyi Lake © Zin Mar Than

Bibliographic information published by the German National Library:
The German National Library lists this publication in the Deutsche Nationalbibliografie;
detailed bibliographic information is available at <http://dnb.d-nb.de>.

© Franz Steiner Verlag, Stuttgart 2017
Print: Hubert & Co, Göttingen
Printed on acid-free and age-resistant paper.
Printed in Germany.
ISBN 978-3-515-11796-8 (Print)
ISBN 978-3-515-11797-5 (E-Book)

CONTENTS

LIST OF TABLES

LIST OF FIGURES

LIST OF ABBREVIATIONS

ASEAN	=	Association of Southeast Asian Nations
CDMA	=	Code Division Multiple Access (channel access method in cellular networks)
cf.	=	confer, compare
DAAD	=	Deutscher Akademischer Austauschdienst (German Academic Exchange Service)
e.g.	=	exampli gratia (for example)
FFI	=	Flora and Fauna International (an NGO)
FOW	=	Friend of Wildlife (an NGO)
GDP	=	Gross Domestic Product
GSM	=	Global System for Mobile Communications (channel access method in cellular networks)
HIV	=	Human Immunodeficiency Virus
IUCN	=	International Union for Conservation of Nature and Natural Resources
KIA	=	Kachin Independence Army
KIO	=	Kachin Independence Organization
KNU	=	Karen National Union
MAXQDA	=	Software package for qualitative data analysis
NCA	=	Nationwide Ceasefire Agreement
NGO	=	Non-governmental Organization
NLD	=	National League for Democracy
R&D	=	Research and development
Ramsar	=	Convention on Wetlands of International Importance (called after Ramsar, Iran, where the convention was signed)
SIM	=	Subscriber Identification Module (integrated circuit chip for mobile phones)
SME	=	Small and medium size enterprises
SPSS	=	Software package for statistical analysis
SWOT	=	Acronym of Strengths, Weaknesses, Opportunities and Threats
UN	=	United Nations
UNESCO	=	United Nations Educational, Scientific and Cultural Organization
UNICEF	=	United Nations International Children's Emergency Fund
UWSA	=	United Wa State Army
WCED	=	World Commission on Environment and Development

PREFACE

This study was submitted as a doctoral thesis under the title "Socio-Economic Development Potentials in the Indawgyi Lake Area, Kachin State, Myanmar" to the Faculty of Mathematics and Natural Sciences of the University of Cologne. The date of the thesis defense was on 24th October 2016.

Prof. Dr. Frauke Kraas and Prof. Dr. Josef Nipper were the reviewers.

ACKNOWLEDGEMENTS

This research was possible because of the encouragement and generous cooperation of my doctoral advisor, Prof. Dr. Frauke Kraas, who has supported me personally throughout my academic career, and has put her trust into my research and personal abilities.

I also like to thank Prof. Dr. Josef Nipper for taking his time and for his effort to accompany fieldwork together with Prof. Dr. Frauke Kraas. In particular, his expertise in quantitative methods was helpful. Special thanks go to Dr. Regine Spohner for her help in developing maps and figures.

I also like to express my gratitude to the generous cooperation of experts and citizens of Indawgyi Lake, namely the Mohnyin General Administrative Office, members of the Rural Development Committee, Principals Dr. Aung Min Win and Dr. Kyaw Tun and their teams from Mohnyin Degree College, Indawgyi Wildlife Conservation team, Friends of Wildlife team in Indawgyi, Mohnyin Fishery Department, Myitkyina Environmental Department, Myitkyina Forest Department, Myitgyina KIO office, Myitkyina University, Myitkyina Hospital, Myitkyina Basic Education Department, Myitkyina Agriculture Department, Myitkyina Hotel and Tourism office, Yangon Fauna and Flora International office and Ministry of Environmental Conservation and Forestry, Naw Pyi Taw.

Taking place in the Indawgyi Lake Area, the data collection processes would not have been possible without the support of Prof. Dr. Frauke Kraas and her colleagues Dr. Seng Aung and Daw Zawng Nyoi from the Geography Department of Myitkyina University. Thin Le War, my research assistant from Mamomkai village, who studies at Myitkyina University, helped not only to carry out questionnaires, but also gave me deeper insight into life in the area. Such insights were also shared with me by Daw Khaine Khaine Swe and her team from the Friends of Wildlife in Indawgyi.

I also have to extend my gratitude to the working group (AG Kraas) Birte Rafflenbeul, Dr. Marie Pahl, Dr. Judith Bopp, Tine Trumpp, Franziska Krachten, Gerrit Peters, Benjamin Casper, and Li Zifeng, who provided a vibrant and comfortable working environment and gave the time of discussion during writing this thesis in Germany. Birte, Marie, Dr. Pamela Hartmann and Lothar Kinzelmann, a former colleague of the Deutsche Welthungerhilfe, were also engaged in proofreading; I appreciate their help very much. I also owe thanks to Dr. Alex Follmann, who spent time for discussion and proofreading.

I am very grateful to Dr. Johannes Müller and Mr. Karl-Heinz Korn from the International Student Office of the University of Cologne for support and taking me on a scholar life at the University of Cologne.

Doing this research over about three years would not have been possible without the generous PhD grant (from 2013–2016) given by the German Academic Exchange Service (DAAD).

My sincere thanks go to Mrs. Christel Kraas for generously accommodating me while waiting for a dormitory in Cologne. In particular, I also owe thanks to many people living in the Indawgyi Lake Area for helping me doing my research work and for just their hospitality.

Finally, a big thank goes to people in Germany and Myanmar and to my family and friends, who supported me in one way or another before and during the process of writing this thesis.

SUMMARY

Indawgyi Lake in Kachin State, located in the northern part of Myanmar, is the largest inland freshwater lake of the country with a rich aquatic flora and fauna, rich biodiversity and is still a largely intact ecosystem. Since 1999, the area has been declared as the "Indawgyi Wetland Bird Sanctuary". Around the lake area eleven village tracts with 38 villages consists with a little more than 50,000 population. However, the socio-economic development of the area is hampered by the protracted unstable political situation and by factors caused by its peripheral location. These factors resulted in poor administrative structures and infrastructure deficits. At the same time the local population, depending heavily on the natural resources, is facing mounting challenges, some of which are internal, others also externally influenced, such as an increasing population (partly due to in-migration) and the rising exploitation of natural resources. Deficits in handling environmental issues are evident. In general, research on the potentials of the area is absent.

These facts and the growing importance of a sustainable development in today's globalized world have led to apply the concept of endogenous development as intellectual base for this study. The research aim – finding out the potentials of the area and discussing possible development paths – can be subdivided into three objectives: 1) investigating the present socio-economic conditions (demographic, social, infrastructure, economic, governance and conservation aspects) of the Indawgyi Lake Area; 2) investigating how the local people evaluate the current situation and the future development; 3) based on the results, identifying and discussing the potentials of the region for future development in detail. Especially solutions for current critical aspects and future threats are discussed, as well as ideas for ways and means to develop eco-tourism in the area.

The empirical work was conducted in two phases in 2014 and 2015 using a mixed method approach. The quantitative primary data were collected applying a questionnaire for a total of 216 households in ten villages around the Indawgyi Lake. These data were analysed using SPSS. To get qualitative data altogether 54 experts were interviewed and field observation and participant observation methods were applied. The interview data were analysed using MAXQDA. The results of the quantitative and qualitative analyses were combined by triangulation and interpreted based on the SWOT concept.

Main economic activities are agriculture, fishery and gold mining. One of the current challenges of the agricultural sector is a sub-optimal land use management (e.g. monoculture cropping systems). Challenges of the fishery sector are overfishing, ignoring the closed season and practicing illegal fishing methods. the In the gold mining sector visible and invisible impacts can be mentioned as current and future threats. For instance, sedimentation can be observed in the lake. Other visible threats like drug use and health problems are predominant in the mining area. As an invisible and a future threat mercury contamination can be pointed out, because mercury is used in gold extraction.

Imbalances in the migration patterns, which have negative impacts on development, can be found in the area. The area offers unskilled labour opportunities in the mining and fishery sectors. Simultaneously people, who are educated, have to leave the area and look for qualified jobs in other parts of Myanmar resulting in a brain drain for the region. Currently, a big threat for the area is the unstable political situation including ethnic armed conflicts. Peace negotiations between ethnic armed groups and central government are underway since 2012.

The identified challenges are discussed sector by sector and possible solutions are described. For example, how adverse effects of the monoculture cropping system can be mitigated, how overfishing can be prevented or how the brain drain process can be reduced. The economic development potentials are identified and discussed as for instance added value processes for farming, fishery and mining products, which are still underdeveloped due to lacks in infrastructure. Since 2013 upgrading of the main road has started and in 2015 electricity supply was established already for several villages. These are positive signs for future development. In particular, a focus is put on eco-tourism, which has not developed until now, as a fourth basic economic sector for the area.

However, also possible negative impacts of development need to be kept in mind. For instance, a better accessibility includes the threat, that the carrying capacity of the ecosystem of the area will be overstrained. Therefore, development of eco-tourism should be environmentally sound. However, the most important factor is to achieve a stable political situation, otherwise the development of the region cannot be moved forward successfully.

ZUSAMMENFASSUNG

Der Indawgyi Lake, im Kachin State im nördlichen Teil Myanmars, ist der größte Süßwassersee des Landes mit einer reichen aquatischen Flora und Fauna, einer hohen Biodiversität und einem (weitgehend) intakten Ökosystem. Seit 1999 ist die Region als Schutzgebiet („Indawgyi Wetland Bird Sanctuary") ausgewiesen. Die sozio-ökonomische Entwicklung wird allerdings eingeschränkt durch eine schon lang andauernde instabile politische Situation und durch Faktoren, hervorgerufen durch die periphere Lage. Mangelhafte Verwaltungsstrukturen und Defizite in der Infrastruktur sind als Folgen zu nennen. Gleichzeitig ist die lokale Bevölkerung – stark abhängig von den natürlichen Ressourcen – steigenden Herausforderungen ausgesetzt. Einige sind lokal verursacht, andere kommen von außen wie z.B. ein Bevölkerungswachstum (z.T. Einwanderung) und eine wachsenden Ausbeutung natürlichen Ressource. Defizite im Umgang mit der Umwelt sind evident. Insgesamt liegen kaum wissenschaftliche Informationen über die Region vor.

Diese Situation und die wachsende Bedeutung nachhaltiger Entwicklung in einer heute globalisierten Welt sind der Anlass, für diese Studie das Konzept endogener Entwicklung als gedankliche Basis zu nehmen. Das Ziel der Untersuchung – Identifizierung und Diskussion der Potentiale – ist unterteilt in: 1) Analyse der aktuellen sozio-ökonomischen Struktur (demographische, soziale, infrastrukturelle, ökonomische, administrative Aspekte sowie Umweltschutz) der Region; 2) Aufzeigen der Einschätzungen der lokalen Bevölkerung zur gegenwärtigen und zukünftigen sozio-ökonomischen Situation; 3) Herausarbeitung der Potentiale der Region für zukünftige Entwicklung, basierend auf den gefundenen Resultate. Insbesondere werden Lösungen für gegenwärtige kritische Aspekte und zukünftige Gefahren angesprochen und Ideen für einen Ökotourismus diskutiert.

Die empirische Arbeit wurde in zwei Phasen in 2014 und in 2015, basierend auf einem „mixed method"-Ansatz, durchgeführt. Primärdaten wurden durch eine Haushaltsbefragung von 216 Haushalten in zehn in Seenähe gelegenen Orten erhoben und mit SPSS analysiert. 54 Experteninterviews bilden die Grundlage für die qualitative Erhebung, ergänzt durch Informationen aus teilnehmender und allgemeiner Beobachtung. Die Interviews wurden mit MAXQDA analysiert. Alle Daten (quantitative und qualitative) wurden durch Triangulation miteinander verknüpft und vor dem Hintergrund des SWOT-Konzeptes interpretiert.

Die Hauptwirtschaftsbereiche sind Landwirtschaft, Fischerei und Goldgewinnung. Eine der aktuellen Herausforderungen in der Landwirtschaft ist eine suboptimale Anbaupraxis (z.B. Mono-Anbau-System). Probleme in der Fischerei sind Überfischung, die Nichtbeachtung von Fangverbotszeiten und die Anwendung illegaler Fangmethoden. Die vom Goldbergbau hervorgerufenen Gefahren lassen sich aufteilen in sichtbare und unsichtbare. Z.B. fließen Abwässer in den See und erzeugen Sedimentation. Andere Gefahren wie Drogenkonsum und Gesundheitsprobleme sind im Bergbaugebiet weit verbreitet. Als unsichtbare und zukünftige

Gefahr ist zudem die Kontaminierung mit Quecksilber, welches bei der Aufberei-
tung des Goldes eingesetzt wird, zu nennen.

In der Region werden durch Migration starke Ungleichgewichte erzeugt. Die
Region bietet Möglichkeiten für ungelernte Arbeitskräfte im Bergbau und in der
Fischerei. Gleichzeitig verlassen gut ausgebildete junge Menschen die Region und
suchen sich Arbeitsstellen in anderen Teilen Myanmars und tragen so zu einem
brain-drain für die Region bei. Gegenwärtig ist auch die instabile politische Situa-
tion mit bewaffneten ethnisch motivierten Auseinandersetzungen eine große
Gefahr. Friedensgespräche zwischen den bewaffneten ethnischen Gruppen und der
Zentralregierung sind seit 2012 wieder aufgenommen worden.

Die Herausforderungen werden diskutiert und Lösungswege werden aufge-
zeigt. Wie kann z.B. das Mono-Anbau-System reduziert oder das Überfischen
vermieden werden oder wie lässt sich der brain-drain-Prozess abmildern. Die
ökonomischen Entwicklungspotentiale werden identifiziert und u.a. diskutiert wie
eine Weiterverarbeitung der erzeugten Produkte in der Region erfolgen kann, was
bisher kaum geschieht wegen Mängel in der Infrastruktur. Seit 2013 wird die
Hauptstraße, die um den See führt, ausgebaut und 2015 sind mehrere Dörfer an das
öffentliche Stromnetz angeschlossen worden, gute Zeichen für die zukünftige
Entwicklung. Insbesondere wird in der Studie auf die Etablierung eines Ökotourismus, der bisher nicht vorhanden ist, als ein möglicher vierter Wirtschaftssektor für
die Region eingegangen.

Aber es sind auch negative Entwicklungseinflüsse zu berücksichtigen. So
besteht die Gefahr, dass durch die angestrebte bessere Erreichbarkeit der Region
die Aufnahmekapazität des Ökosystems überschritten wird. Insofern ist ein zukünf-
tiger Ökotourismus in enger Abstimmung mit den Belangen des Umweltschutzes
zu gestalten. Am wichtigsten allerdings ist es eine stabile politische Situation;
ansonsten kann die Entwicklung der Region kaum erfolgreich sein.

1   INTRODUCTION

1.1   THE AIM OF THE STUDY

For several years – in particular since 2016, when a new government took over power after the election in 2015 – an intensive transformation process has been started in Myanmar. It also has to be pointed out, that in Myanmar many periphery regions do exist, which are lagging behind distinctly. For these regions often only limited information about the structures and development potential are available, but such information is vital to meet the transformation and to have a successful regional development. This also holds true for the Indawgyi Lake Area in the northern part of the country.

The above described situation is the starting point for carrying out the following study on the Indawgyi Lake Area. The aim of the research – finding out the potentials of the area and discussing possible development paths – is divided into three objectives:

1. investigating the current socio-economic situation using demographical, social and economic indicators,
2. investigating how the local people evaluate the current socio-economic situation and future development and
3. identifying and discussing the region's sustainable development potentials.

1.2   MYANMAR UNDER TRANSFORMATION

After the Second World War during the first years of independence between 1950 and early 1958 Myanmar was one of the economically most advanced countries in Southeast Asia. In particular, Myanmar exported agricultural products (e.g. rice) and offered good higher education opportunities to students from other countries in Southeast Asia (e.g. South Korea and Singapore) (Hauff 2009: 1). However, in late 1958 the situation changed because the ruling party of the democratic government split into two factions and various ethnic groups began to demand autonomy[1] (Bünte 2011). As a result, the military took power in a coup in 1962 and appointed a new government cabinet, which was made up of selected high-ranking generals

---

[1]   According to the Panglong agreement they were entitled to do so after 10 years.
The Panglong agreement is an agreement, which has been signed in Panglong between the Burmese government under General Aung San and the Shan, Kachin, and Chin in 1947. The agreement accepts "[...] Full autonomy in internal administration for the Frontier Areas" (Tinker 1984: 404–405, Ethnic National Council of Burma 1947).

(Huang 2012, Bünte 2011, Hauff 2009: 9–10, Nakanishi 2007). In 1974, this military government adopted a new constitution after abolishing the 1947 constitution and dissolving the parliament (Hnin Yi 2014).

Thereafter the 'Burmese Way to Socialism' was launched (Huang 2012, Maung Aung Myoe 2009: 60, The Revolutionary Council of the Union of Burma 1962) and Myanmar moved toward self-imposed isolation (Thant Myint-U 2009). As a result the economy deteriorated rapidly despite its richness in natural resources, including oil and gas, gems and teak (Bünte 2011, Tyn Myint-U 2010: 17). Under the socialist regime all economic sectors (e.g. international trade, banks), all social services (e.g. health, education) and all infrastructure sectors (e.g. transportation, communication) were nationalized (Hnin Yi 2014, Bünte 2011, Maung Aung Myoe 2009: 61). In particular, farmers suffered from this system because they had to sell their products to the state below market prices. In 1974, the 'Green Revolution' was initiated in the agricultural sector using high-yield varieties and applying of agrochemicals in order to improve efficiency and productivity (Thein 2004: 5). However, this was not a sustainable solution to overcome the economic problems and to widely increase productivity. The decline in exports led to significant reduction of imports and public investment, thus additional money was printed in order to reduce the national budget deficit. This finally resulted in a significant increase in inflation and a severe economic crisis at the end of the 1980s (Bünte 2011).

In 1988 after 26 years, the socialist system came to an end (Taylor 2012), a new military junta abolished the 1974 constitution, took power and promised to hand over the power after holding a multiparty national election (Hnin Yi 2014). A market-oriented economy was officially adopted in March 1989 (Hnin Yi 2014, Taylor 2012, Bünte 2011, Thein 2004: 6). However, the military regime did not acknowledge the result of the 1990 election and maintained power continuously (Data Team 2016, Bünte 2011, European Union n.d.–a). The regime was ideologically convinced that only the military could serve the interests of the nation and also claimed that a new constitution needs still to be drafted (Hnin Yi 2014, Nakanishi 2013). As a result, the United States and the European Union stepped up the economic sanctions against the military regime (Taylor 2012, Bünte 2011). Consequently, most of the foreign investment came from the neighbouring countries, namely China, Thailand, India, Malaysia, Singapore and South Korea (Taylor 2012, Tyn Myint-U 2010: 17). However, the government paid little attention to the economic sanctions (Taylor 2012). It developed the agricultural sector and intensified its effort to build infrastructures across the country (Hnin Yi 2014, Bünte 2011). Therefore, within a few years, an incredible number of bridges, dams, reservoirs and irrigation canals were constructed. A new rice trading policy was adopted, which delegated the authority to purchase rice for food security to the government and allowed farmers the right to market their products freely (domestically or internationally) (Nakanishi 2013, Tyn Myint-U 2010: 18). In addition, education infrastructure (schools, colleges and universities) and health infrastructure (health care centres, hospitals, medical clinics) were established (Hnin Yi 2014, Bünte 2011). Unfortunately, due to insufficient availability of funds and poor organisation, the infrastructure built was not fully equipped (e.g. libraries lacked books, vocational trainings was insufficient because of lack of human resources).

Nevertheless, between 1988 and 2010 the private business sector had grown but government/semi-government organizations still played the dominant role in the economy and SMEs (Small and Medium Enterprises) could not compete with the big government-backed corporations (Nakanishi 2013, Bünte 2011). Characteristic features of crony capitalism[2] were present, with only a handful of businessmen benefitting from the market-oriented economy under strict government control (Bünte 2011). The business environment was not conducive to development with no proper banking facilities in place, while exchange rate fluctuations in the prevailing black market made normal business operations all but impossible.

In 2008, the new constitution drafted by the military regime was approved by a highly unfair public referendum and subsequently the national election was held in November 2010 (Huang 2012). As a result, in March 2011 Myanmar's long ruling military government handed over power to a new civilian government, which consisted exclusively of members of the military-backed Union Solidarity and Development Party (USDP) (Taylor 2012, Huang 2012, Bünte 2011). All administrative and legislative bodies at the central, regional and local levels were also controlled by this military-backed party (Huang 2012, Bünte 2011). Therefore, it did not seem that real change could happen easily (Nakanishi 2013). With a constitution under which 25% of the parliament seats are reserved for the military, a process towards to introducing democracy and a more transparent and equitable society had begun (Robinson 2014, Gaens 2013).

Under this new government a reform process with democratic transition, economic and social reforms, and an ethnic peace process, initiated by the president himself, took place (Bünte and Dosch 2015, Nakanishi 2013, Gaens 2013,). The government stated that the reform process would be inclusive and participatory, bringing all interested parties together, including civil society and the private sector (Bünte and Dosch 2015, Nakanishi 2013, European Union n.d.–b).

The milestone of the peaceful political transition was the release of opposition leader Daw Aung San Suu Kyi from house arrest along with freeing a substantial number of political prisoners (Bünte and Dosch 2015, Nakanishi 2013, Effner 2013, Effner and Schulz 2012). In addition, in August 2011 the government publicly reconciled with ethnic armed groups and invited them "to secure lasting peace" in the country (Nakanishi 2013, Min Zaw Oo 2014). Furthermore, in September 2011 the president suspended the controversial Myitsone Dam Project in response to the rejection of the project[3] by civil society groups (Bünte and Dosch 2015, Effner and Schulz 2012, internationalrivers n.d). Myanmar people began to support the idea that new Myanmar government should be given a chance to prove itself different from its predecessor. This situation led people to examine more in-depth how the emergence of the new government has affected Myanmar politics, the economy,

---

2   Crony capitalism is a system in which some businessmen and entrepreneurs, who have intensive relations to the government and control most of the important sectors of economy.

3   Jointly developed with Chinese construction companies for hydro-power supply.

social affairs and foreign relations and how the recent political changes have influenced the relations between the government and the opposition.

With the 2012 by-election the major opposition party, the National League for Democracy, came into parliament with 43 seats (Inter-Parliamentary Union 2016, Bünte and Dosch 2015, Nakanishi 2013, Effner 2013). And by mid 2012 a number of ethnic armed groups signed bilateral ceasefire agreements with the government across the country (Min Zaw Oo 2014, Effner 2013, Nakanishi 2013). During 2012 several draft laws (e.g. environmental, foreign investment) were drawn up. The reform process showed to be genuine. Consequently, in April 2012 the EU's sanctions imposed on the government were suspended (Effner 2013, Taylor 2012) and then lifted in 2013 (with the exception of the arms embargo) as a mean to welcome and encourage the reform process (Taylor 2012, European Union, n.d.–b). The United States also recognized the progress and encouraged the government's reform process by easing its economic sanctions on Myanmar (Min Zaw Oo 2014, Nakanishi 2013). Simultaneously, Japan decided to resume Yen loans to Myanmar, after cancelling the old debts (Nakanishi 2013). This is in line with the attempts of the president, Thein Sein, to overcome the formerly unbalanced policies (e.g. excessive relation to China (Thant Myint-U 2009)) and to reach a balance in foreign policy (China-U.S.) and in domestic policy as well (Nakanishi 2013).

Of course, there was criticism of the government such as:
1. Still ignoring the human rights, for example in 2014, peaceful education protests were brutally disbanded in Letpadan, Bago Region, and about 69 students and activists remained in detention (Ye Mon and Verbruggen 2016).
2. Tensions between government and several ethnic armed groups, which have not signed ceasefire agreements, are still high due to lack of trust (Wa Lone 2016, Nakanishi 2013).
3. Still increasing crony capitalism interrelated with all levels of corruption (Htoo Thant 2016).

Nevertheless, Myanmar held the chairmanship of the Association of South East Asian Nations (ASEAN) in 2014. This had never happened since 1997, when Myanmar became a member of ASEAN. It shows that Myanmar now got a full place in the association and is internationally accepted. Additionally, in late 2015, the launch of the ASEAN Economic Community (AEC) was held in Myanmar. On this meeting a regional trade and cooperation agreement was signed, which offers the prospect of further economic development (Robinson 2014).

In the November 2015 election the opposition party, the National League for Democracy (NLD), won by a landslide victory (over 80% of the elected seats in the national parliament; however, 25% of parliament seats are reserved for the military and only 75% of parliament seats are elected according to the 2008 constitution), because the people wanted a real change (Pedersen 2016, Bärwaldt 2016). Consequently, the NLD got to choose the president, who is elected by an electoral body made of three separate committees, as well as the head of the Supreme Court and the chief ministers of each of the 14 states and regions. In March 2016 the new president was elected. The ministries were reorganized from previously 36 ministries with 96 ministers and deputy ministers to now 21 ministries with 21 ministers to cut the costs and reduce the budget deficit (Ei Ei Toe Lwin 2016). However, according to the 2008 constitution the president must accept that the commander-in-chief will appoint three ministers for three ministries (e.g. ministries of defence, home affairs and border affairs) (Pedersen 2016). Only one new ministry was created for ethnic affairs, which reflects the priority that the new government, gives to the peace and "national reconciliation" process (Ei Ei Toe Lwin 2016). The new government officially took power on 1st of April 2016.

Figure 1.1: Spatial administrative structure of Myanmar and Kachin state

The spatial administrative structure in Myanmar consists of six levels namely national, state/region, district, township, village tract (sub-township), and village (see figure 1.1). As mentioned before the administrative units below the national level had hardly any autonomous decision-making power in the past. This will be changed under the new government and by doing so, Myanmar is getting closer to a federal system.

In general, the present and future situation of 'Myanmar under transformation' can be summarized as follows:
1. change into a democratic society with a federal system,
2. embrace ethnic participation in political processes,
3. reorganised and open economic system.

1.3   THE PERIPHERAL SITUATION OF RURAL AREAS IN MYANMAR

In Myanmar the states and regions have started to develop since the turn of the century. But the development of Kachin state is lagging behind and will be taken in this subchapter as an example for the situation of peripheral areas, because the situation in Kachin state is quite characteristic as a periphery.

The Kachin state is – as the Shan and Rakhine state, too – an area where civil war based on ethnic conflicts still takes place. The insecurity and instability caused by this situation results to a large extent in a lack of powerful law and control, either exerted by the national government or the ethnic group organisation. This lack causes firstly, illegal resource exploitation with the consequence of environmental degradation (e.g. jade mine disaster in Hpakant), secondly, illegal trading supported by the lack of efficient control mechanism and thirdly, a weak and not well-developed infrastructure caused by the unstable political situation. As a consequence, the reaction of people results quite often in out-migration, which can be characterised as follows: well-educated people from the area out-migrated because they want to go to central Myanmar such as Yangon or Mandalay, where they expect to get more opportunities. This out-migration leads to a brain drain. Simultaneously, these peripheral areas are faced with huge immigration. People from all over Myanmar come for labour in gold and jade mining, and other activities like fishing or logging. They come because they see their economic opportunities. However, the educational level of these immigrants is quite low. These opposite migration schemes are a big challenge for regional development, because the newcomers have little knowledge, the others who know more move away. The result is that the local knowledge is weakened.

These problems come on top of the big challenges of the transformation. Insofar a good regional development policy is necessary. A fundamental element for that is to strengthen the knowledge of the people living in the area. So all stakeholders should be aware of that and try to find a solution. Without any doubt the on-going civil war has to be stopped, otherwise a successful development path cannot be implemented.

1.4 AN INTRODUCTION INTO THE RESEARCH AREA

1.4.1 Physical geographical pattern

Topography

Indawgyi Lake in Kachin State is the largest inland freshwater lake of Myanmar, located in the northern part of Myanmar (see figure 1.2). The lake is situated at an altitude of about 170 m and stretches over a distance of 23.1 km from north to south; 7.5 km from east to west, and is covering an open water area of 120 km² (Davies et al. 2004: 222)[4]. The basin is asymmetrical with a depth of between 15.9 m to 22.2 m and it is surrounded by densely forested mountain ranges with altitudes of between 200 and 1,300 m (Arino et al. 2009, Jarvis et al. 2008). The lake's catchment area stretches over 825 km²[25] (Davies et al. 2004: 223). The lake is fed by several streams at the northeast end. Here, the Indaw stream flows into Moegaung stream, which feeds the Ayeyarwady River.

Climate and soil

The area has a tropical wet and dry or savanna climate (Rubel and Kottek 2010, Chen and Chen 2013) with an annual rainfall of about 2,000 mm with the mean relative humidity of 80% to 90%. According to the information of the Meteorology and Hydrology Department, Mohnyin Station, Mandalay (2010) the average monthly temperature is varying between 17°C and 28°C. It can however stretch from as low as 4.6°C during December and January to a peak of up 40°C in April. Early morning mist is common during the cold season (Davies et al. 2004: 222).

The soils in the lower parts are rich and thus very suitable for agricultural use: meadow alluvial soils are found around the lake area and in the plains. They are mainly composed of silt and clay (70% and 20% respectively), have a pH-value around 6 and the ratio of nitrogen and carbon is 7:13. This soil is suitable for rice cultivation, sugarcane, groundnut, beans and vegetables (Naing Naing Latt et al. 2010). Meadow swampy or grey soils are found particularly in wet areas with poor drainage, especially north of the lake. The lower ground layers show blue or grey colour with tiny red or brown spots. The soil is sticky and has high contents of clay (gley). The pH value ranges around 6.5, that is, it is slightly acid. Red and yellow brown forest soils predominate in the mountain ranges and these are less suitable for agricultural purposes as they are particularly prone to erosion (Ministry of Agriculture and Irrigation 2002).

---

4   775 square km is the total protected area size of the Indawgyi Wetland Bird Sanctuary, which consists of the water body (120 square km and large parts of the catchment area).

5   825 square km is the total catchment area, but not the entire area is under protection (Davies et al. 2004: 223).

Figure 1.2: Map of the research area

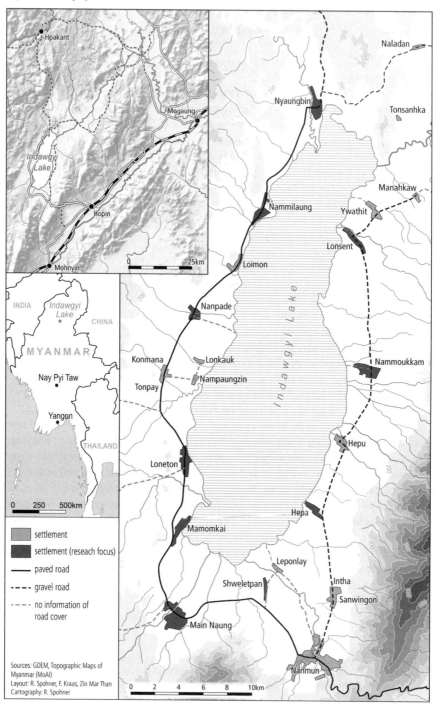

Natural zones and land use

Five different land use zones can be distinguished (Kraas and Zin Mar Than 2016). (1) The alluvial plains around the lake which are mostly covered with rice fields. (2) Seasonally inundated and waterlogged plains, which are covered with herbaceous marsh, scrub swamp and swamp forest. (3) In the open water of the lake, especially at the northern end between Nyaungbin and the outflow of Indaw stream as well as at the southern end around Nanyinkha stream are extensive areas of herbaceous marches and water hyacinths. (4) As the water has a relatively high transparency (up to 3.5 m), extensive beds of submerged and floating leaved macrophytes can be found in some places (Davies et al. 2004). (5) The mountain ridges are mostly covered with broad leaf forests with many teak trees (*Tectona grandis*).

In accordance with the natural potentials, different agricultural zones can be identified: (1) Le lands are mostly found in the Indawgyi plain. They are broken up into many small units. Here, farmers cultivate rain-fed paddy, with yields corresponding with the rainfall. (2) Ya lands are found along the streams and on cleared forestland. Here, maize, groundnut, sesame and mustard are grown. (3) In garden lands, including home, vegetable and fruit gardens, a large variety of subsistence agricultural products are grown, like mangoes, citrus fruits, flowers and vegetables. (4) Taungya lands are found on the mountain slopes where only a thin soil cover allows two to three years of cultivation after clearing and burning (slash-and-burn cultivation). Here, upland rice, maize, sesame, pulses and vegetables can be grown.

Biodiversity: Since 1999, the Indawgyi Lake Area is under conservation as the 'Indawgyi Wildlife Bird Sanctuary', managed by the Nature and Wildlife Conservation Division of the Forest Department. Its rich biodiversity makes it a unique site: Altogether 64 fish species are recorded in the lake basin, inflowing streams and marshy areas. Among them, three species are endemic (Davies et al. 2004: 226). The lake is one of the most important bird refuges in Southeast Asia, mostly migratory birds; 95 species of water birds have been recorded (Nyo Nyo Aung 2008: 20). Numerous famous migratory and resident species can be found, such as the Pallas's Fish Eagle (*Haliaeetus leucoryphus)*, the Sarus Crane (*Antigone antigone)*, the Lesser Adjutant Stark (*Leptoptilos javanicus)*, the Purple Swamphen (*Porphyrio porphyrio)*, the Greylag Goose (*Anser anser)* or the Oriental Darter (*Anhinga melanogaster)* (BirdLife International 2015). Among them 10 endangered species are, such as the 'critically endangered' winter-staying Baer's Pochard (*Aythya baeri)* and the non-breeding White-rumped Vulture (*Gyps bengalensis)*: the 'endangered' resident Green Peafowl (*Pavo muticus)* and the 'near threatened' non-breeding Spot-billed Pelican (*Pelecanus philippensis* (BirdLife International 2015). The Lesser Adjutant Stark, Baer's Pochard, Spot-billed Pelican and Sarus Crane are on the Myanmar list of the 'Totally Protected Animals', protected by law (State Law and Order Restoration Council 1994; Kyaw Nyunt Lwin and Khin Ma Ma Thwin 2003: 150). Also the highest number of the Eastern Hoolock gibbons is recorded in the area (Geissmann et al. 2010).

With this rich biodiversity and intact ecosystem, the Indawgyi Lake Wildlife Sanctuary belongs to the ASEAN Heritage sites. In February 2014 the government submitted an application to the UNESCO to nominate the area as World Natural Heritage; at present the area is on the tentative list. Since February 2016 the area was already listed as a Ramsar[6] Site. Even though a management plan exists, it is, according to BirdLife International (2015), out-of-date and not comprehensive.

1.4.2 Social economic situation of the Indawgyi Lake Area

While the physical and environmental situation of the Indawgyi Lake Area have been described in a few publications (as summarized above), knowledge of the current socio-economic and the political situation of the area is very minor. Improving this knowledge is part of the aim of this research. In this section only some general features of the socio-economic situation will be presented.

The Indawgyi Lake Area (see figure 1.2) consists of eleven village tracts with 38 villages stretching over an area of 1,211.4 km². With about 8,758 households and a population of 50,014, the Indawgyi Lake Area is already quite densely populated with 41.3 person per km² (compared to Kachin State with 18.9 per km²) (Department of Population 2015, Kraas and Spohner 2015).

The Indawgyi Lake Area belongs to the Mohnyin Township (Mohnyin District). The area is located in a distance of about 180 km (112 miles) southwest from Myitkyina, the capital of the Kachin State (see figure 1.1). The nearest town, Hopin, and its administrative town, Monhyin, are 24 miles and 33 miles in distance from Loneton, the main village at the west bank of the lake (see figure 1.2). These two towns are gates to have access to the lake. The lake can be reached in a five-and-a-half-hour drive from Myitkyina. The two-lane highway starts from Myitkyina and passes through Namte, Moegaung, Sarhmaw, Hopin, Nanmun and Loneton. There is also another option: Mandalay-Myitkyina Railway mode. During summer, nine miles can be saved taking the short way between Hopin/Mohnyin and Nanmun.

The local residential population and seasonal migrants depend strongly on the natural resources. Agriculture and fishing together with some mining are the main sectors of the economic base until yet. As a result, some challenges can be found in the research area such as overfishing, illegal mining, and environmental degradation.

One future economic base sector might be eco-tourism because of the nature of the area. This sector has not been developed until yet.

---

6   The Ramsar Convention is dealing with wetlands of international importance. The convention was established in a meeting, which was held in the Iranian city of Ramsar in 1971 and came into force in 1975. It is an intergovernmental treaty that provides the framework for the conservation and wise use of wetlands and their resources (Ramsar 2016).

1.5   STRUCTURE OF THE STUDY

Based on the research aims mentioned in subchapter 1.1 the concept of study is organized in the following way (cf. figure 1.3): It can be subdivided into two parts, a theoretical and an empirical one, both of which support the research aim 'finding out the potentials of the area and discussing possible development paths'. The research area belongs to the rural part of Myanmar and is – as mentioned already a region under conservation. Therefore, ideas of the endogenous development concept have to be combined with the consideration of sustainable development theory to get possible development paths for this rural region. Based on such a theoretical and conceptual framework the empirical part emerges. It consists of two main parts: one is the analysis of the current situation of demography, social structure, infrastructure, economy and governance as well as conservation and the second is directed towards evaluation of current and future situation done by locals and experts. All outcomes are intended to get a good base for creating possible development paths, which is an overall aim of the research.

Figure 1.3: Structure of the study

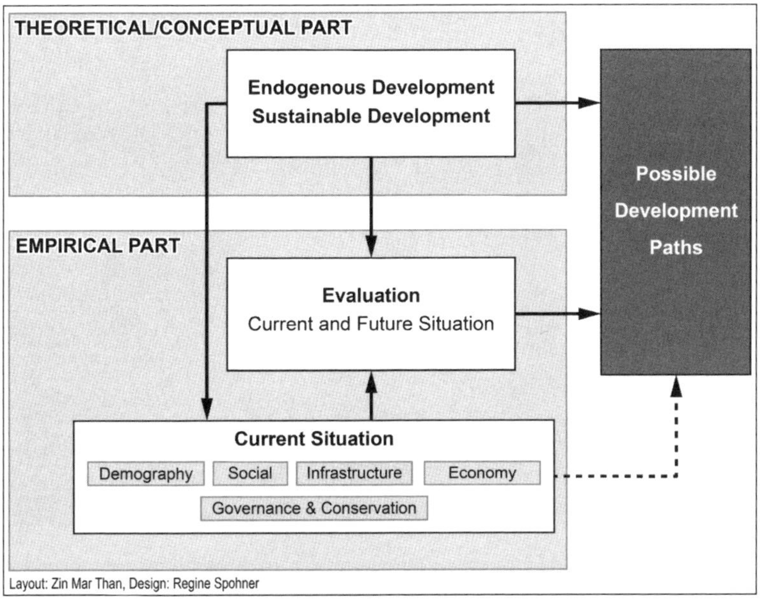

After having already presented in this chapter the purpose of the research, the situation of Myanmar under transformation and the peripheral situation of rural areas in Myanmar as well as having given an introduction into the research region, and the structure of the study, the following chapter 2 presents the theoretical and conceptual framework. It describes the concepts of endogenous development and sustainable development, and discusses how to facilitate these two concepts as well as how to bring this idea to this research. Finally, the key research questions are deduced. Chapter 3 explains the research design and describes, how the data collection has been carried out and which methods have been used in analysing the data and how the analytical results have been interpreted and evaluated based on SWOT analysis. Included is also a discussion on the ethical consideration related to the fieldwork, which was carried out. Chapter 4 analyses the demographic and social situation of the research area. Chapter 5, 6 and 7 are investigating the situation of the infrastructure, of the economy and of governance and conservation of the area respectively. Chapter 8 presents the evaluations of the current and future situation as seen by the respondents. Chapter 9 discusses the strengths and weaknesses of the region with respect to social development, infrastructure development and economic development. Chapter 10 then aims towards discussing the future development in the region with respect to possible development paths. Finally, chapter 11 states conclusions and final remarks.

2 THEORETICAL AND CONCEPTUAL FRAMEWORK

In order to foster regional development governments have applied different development concepts in different ways. In general, endogenous and exogenous development concepts need to be distinguished. Over the years, it has turned out that endogenous development concepts are probably the most promising ones, in particular if they are combined with the idea of sustainability. In this chapter these concepts, 'endogenous regional development' as well as 'sustainable development', will be discussed in more detail in order to lay out the theoretical and conceptual framework on which the research is based and has been carried out. The discussion is mainly based on the following literature: Pike et al. (2006, 2011), Tödtling (2011), Vazquez-Barquero (1999, 2003) and Stöhr (1981) for endogenous development and Elliot (2006), Rodriguez et al. (2002) and World Commission on Environment and Development (1987) for sustainable development.

This chapter comprises of three subchapters. Subchapter 2.1 introduces the concept of endogenous development, and of sustainable development as well as the interconnection between these concepts. Then subchapter 2.2 explains the rationale for applying this framework to the study of the Indawgyi Lake Region. Against this background, the final subchapter outlines the key research questions.

2.1 ENDOGENOUS AND SUSTAINABLE DEVELOPMENT AS KEY ELEMENTS OF DEVELOPING A REMOTE REGION

2.1.1 From previous top-down to present bottom-up approaches

Previous regional development approaches (in the 1960s and 1970s) for less developed areas have strongly focused on external factors such as trade (exports, imports) or the mobility of capital (firms), labour and technology between regions and countries (Tödtling 2011: 334). Therefore, associated policies were strongly based on factors such as external demand (export base theory, trade theory), the attraction of leading international firms and technology to growth centres (growth pole theory), and the mobility of capital and labour between economically strong and weak regions (neo-classical growth theory) (Tödtling 2011: 334). Such an external development paradigm is called by Stöhr and Taylor (1981: 475) a 'centre-down regional development approach', because it is often organized and carried out by central governments or external agencies.

Stöhr and Taylor (1981: 465–472) argue that a series of critical reviews and studies have shown that this external development paradigm has not really contributed to improve the economic situation of less developed and peripheral regions and countries. As critical remarks to the external regional development paradigm Tödtling (2011: 334–335) points out a variety of critical aspects:

1. There is a strong orientation on external demand and specific comparative advantages of respective regions.
2. Only a small number of regional factors has been mobilised and used in such development policies.
3. Specific natural resources, tourist sites and low-cost labour have been exploited in less developed regions and countries.
4. Other factors and potentials such as qualified labour, specific skills and competences have been hugely ignored.
5. The key strategy was the attraction of external firms and branch plants. This has often benefitted central locations or growth centres in the regions, which led to very few economic spillovers to less developed areas. Often only labour commuting to the centres and the dispersal of branch plants to the remote areas have taken place, but much less input-output linkages and technology transmissions have happened. In addition, higher level functions such as offices and managerial activities, research and development (R&D) and innovation issues frequently have been neglected in the branch plants, which only offered low quality jobs. As a result, the potential of branch plants in less developed regions has depleted. Additionally, due to growing globalization of the economy often new subsidiaries or branch plants have been relocated to other countries (e.g. in South East Asia and in Eastern Europe as well) where lower costs of production such as land and labour are offered.
6. External strategies generally have not raised the entrepreneurial potential and innovation capability of less developed regions to a significant extent.
7. External strategies have not taken into consideration ecological and sustainability aspects, thus the environmental situation in the respective regions has degraded.

Pike et al. (2006: 13) also argue with respect to regional development strategies: "No unique or universal strategy can be applied to every area or region, regardless of the local context". This statement summarizes the experiences with traditional centre-down development policies and their failure caused by their intension to replicate standardised policies in different areas of the world, regardless of the local economic, social, political and institutional conditions. Policies were considered to have success in any specific case, and to have been transferred and implemented almost without changes in different national, regional and local contexts. In this context, Pike et al. (2006: 13) critique that "traditional top-down policies aimed at achieving economic development have tended to be cut from the same cloth".

In agreement with Assche and Hornidge (2015: 17–28) Pike et al. (2006) argue that "[…] development cannot be copied from a recipe deemed universal". That means, one approach might be applicable in one place due to specific background/history, but this does not guarantee achievements anywhere else, where the context and community are different, and therefore local and expert knowledge are vital.

Moreover, Vazquez-Barquero (1999: 80 and 2003: 176) points out that the previous regional development approaches are considered to supply-led policies, and focused either on infrastructure provision or on the attraction of industries and foreign direct investment (cf. Tödtling 2011: 334). The logic behind this approach was that insufficient infrastructure or absence of firms was seen as the root of the problems of many peripheral regions. Thus, local and regional development and employment-creation policies were articulated around the building of motorways, telephone lines, aqueducts, power supply and other investments in infrastructure. On a wider scale, the question has been raised, whether such kind of investment in infrastructure contributes to a sustainable development (see Pike et al. 2006: 13).

In addition, according to the perception from Vazquez-Barquero (1999) and Pike et al. (2006: 14–15) the introduction or attraction of firms to areas with a weak industrial fabric, for example South East Asian countries, has not been successful. One reason is that the policies were often created in the contexts of strong, state-led national development strategy support. In doing so, inadequate local economic and institutional settings acted as a barrier to the creation of networks of local suppliers around the 'imported firms' (cf. Pike et al. 2006: 13). In some areas, weak or deficient education systems and low skills of the workforce became the main barrier for successful development. Poorly suited social and institutional contexts have also been mentioned as possible reasons for the weakness of traditional development policies. Especially traditional top-down and centralised approaches often overlooked or ignored the assets and resources deeply embedded in localities and regions.

Moreover, due to the unbalanced nature of traditional development policies many local firms were driven out of business because of their relatively lower level of competitiveness. Insofar, heavy investment in infrastructure – with little or no emphasis on other development factors such as the support of local firms, the improvement of local human resources or technology transfers and spillovers – has created only imperfect accessibility to markets. As a result, more competitive external firms have benefitted most from greater accessibility to structurally lagging areas, gaining a greater share of markets and driving many local firms out of the markets.

The failure of traditional top-down policies has led to a serious rethinking of local and regional development approaches by scholars and policy observers. As a result, since 1990 a series of innovative, bottom-up local and regional development policies, so called tailor-made approaches, have emerged (Pike et al. 2006: 16). These policies were not based on a consistent new theory, but were defined rather as a counter-thesis to previous regional development approaches (more detailed in 2.1.2) and were shifted subsequently more towards endogenous concepts (Tödtling 2011: 335).

Pike et al. (2006: 106) introduced the 'levelling up'-concept, a new growth-oriented model, which seeks to increase the economic performance of growing regions and under-performing ones by aiming first of all at growth in each region and its national economy. The economic performance of each territory is considered as the key to improved economic outcomes at the local, regional and national level.

Policies following the 'levelling up'-concept aim to overcome the imbalances of regional development and promote self-sustained development of the local economy using local potentials (cf. figure 2.1). This approach is quite contrasted with the traditional regional development model, which is mainly redistributing growth from growing core regions to under-performing regions (Pike et al. 2006: 107). Its policies oriented to improve the balance in the geographical distribution of economic activity and reduce the regional disparities of income and employment through the use of external resources (Vazquez-Barquero 1999: 84–85) (cf. figure 2.1).

Figure 2.1: Traditional development and new growth-orientated model

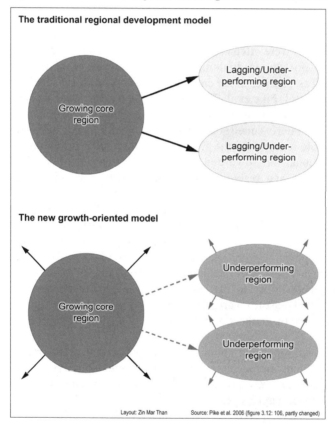

According to Vazquez-Barquero (1999: 83–88, 2003: 146–150), creating a comprehensive and balanced local and regional development strategy based upon a new growth-oriented model is usually based on four axes:
1. the improvement of the competitiveness of local firms,
2. the attraction of inward investment,
3. the enhancement of human capital or labour skills, and
4. the building of infrastructure.

Any of these four axes is integrated into a global strategic framework but with the intention of rooting such economic activities in a certain region, in a way that it meets the idea of using the economic potential of the region properly. Such a comprehensive and balanced strategy means for example, attracting inward investment to an area should be in accordance with the aim of improving the local economic firms, local labour supply and local infrastructure. Similarly, the enhancements in labour skills have to be matched and coordinated with the efforts to boost local firms, to improve infrastructure, and to attract inward investments, etc. Such a balanced and integrated strategy can be achieved only by the systematic participation of local economic, social, and political actors in the planning and development process and by a careful analysis of the economic potential of any area (Vazquez-Barquero 1999: 84–85).

In general, some risks have to be addressed, which are associated with local and regional development approaches (cf. Pike et al. 2006: 20). Probably the main drawback of such a strategy is that it can be extremely time-consuming. For example the development of local and regional coalitions and the coordination of local and regional stakeholders and with other institutional actors need a substantial amount of organizational effort and consumes a considerable amount of time and resources even before the development process starts. Even when the local and regional institutions are established, there is no guarantee of success – neither short-term nor medium-term nor long-term. Apart from the above-mentioned problems, to identify, design or implement the most appropriate development strategy bears a high risk.

Being aware of the critique of the bottom-up approaches, in this research the ideas of an endogenous local and regional development approach will be applied.

2.1.2 Endogenous development theory

Endogenous development theory is inspired by the main paradigms of economic development theories, which from the 1950s to 1970s have dominated theoretical debates (Vazquez-Barquero 2003: 51). Among other theoretical approaches, endogenous development theory retrieves mechanisms from the high theory of development (Krugman 1995) and from transitional growth theory (Lewis 1954, 1958, Fei and Ranis 1964). Both these theories facilitate capital accumulation and growth in a market economy. However, as opposed to these theories, endogenous development theory sees economic growth not like a succession of an equilibrium proposed by neoclassical theory, but like a process characterized by uncertainty and chance (Vazquez-Barquero 2003: 52).

The endogenous development theory also adopts the idea of dependence theory (Amin 1976) that a development approach has to pay attention systematically to economic, social, political and institutional aspects for an organizational structure. But the theory of endogenous development differs from the idea of dependence theory in that the growth of city/region is not decided by its geographical situation or peripheral nature or the level of development within a given period. Endogenous

development theory sees development as a path which depends on the involvement
of natural and human resources and the ability of firms, cities or regions to react
and respond continuously to the challenges of competition at each historical
moment (Vazquez-Barquero 2003: 52).

Furthermore, the endogenous development theory adopts the idea of territorial
theory (Friedmann 1979, Stöhr and Tödtling 1979), which argues that (local)
development agents are the actors who make investment decision and control the
process of change through local initiatives. The territorial development theory is
one of the basic theoretical references of endogenous development theory. Both the
theories have the same concept of economic space and they practice the bottom-up
action in development policy (Vazquez-Barquero 2003: 50). In addition, both
theories understand that each and every territory has its own development potential
that makes a different path to growth. Therefore, each territory has a

> "[…] unique productive system, a labour market, a specific way to organize production, entre-
> preneurial capacity and technological know-how, an endowment of natural resources and
> infrastructure, a social, political and institutional system, tradition and culture through which
> local economic dynamics evolve" (Vazquez-Barquero 2003: 50).

Nevertheless, the position of endogenous development theory differs from the
territorial development theory. In endogenous development theory, the economies
of each city or region are not only integrated into the international relation system,
but rather they are also embedded within the country's system of economic, social,
and institutional relations. Additionally, the local growth path depends on its
provision of resources and local cultural identity (Vazquez-Barquero 2003: 50–52).
Insofar one can agree with Vazquez-Barquero (2003: 44) when he states:

> "[…] endogenous development theory postulates that local development can be articulated
> around any type of activity (agrarian, industrial or services), as long as its production units are
> competitive in the markets".

According to McCombie and Thirlwall (1997: 2–16) one critique to the endogenous
theory is that the theory focuses on the supply-side paying overall little attention to
the demand-side issues of exports, balance of payments constraints on employment
and productivity. Pike et al. (2006: 107) add two further critical aspects:
1. Endogenous theory remains connected to the standard neo-classical assump-
    tions about economically rational agents, fully knowledgeable of alternative
    choices and the consequences of their decisions.
2. Endogenous theory can also be poor in addressing the (historical) social and
    institutional context shaped by geography and place that develop the operation
    of economic growth processes.

2.1.3 Endogenous development approaches

The endogenous regional development approach has been inspired by theories of
economic development, evolutionary economics, innovation, and learning theories
(Tödtling 2011: 340). It was introduced in the late 1970 and since then it has become

popular in developed and developing countries (Tödtling 2011: 333). Important elements of such an approach are for example regional specificities and uniqueness, identity as a source of competitive advantages, regional institutions or social capital and networks as constituent parts (Tödtling 2011: 340). Major roles for development play entrepreneurship and innovation, modern information and communication technologies and the idea that local and regional processes are based on localised tacit knowledge and its exchange (Stimson et al. 2009: 11–12, Tödtling 2011: 337–340). Thus, an in-depth understanding of the regional setting and situation contributes to enhance the potentials for socio-economic improvement.

According to Tödtling (2011: 333) an endogenous approach is more broadly defined in comparison with an indigenous approach, which following Pike et al. (2006, 155) "[…] is based on naturally occurring and/or socially produced sources of economic potential growing from within localities and regions" (Pike et al. 2006: 155). That is, home-grown assets and resources are more locally and regionally nurtured and more committed and more durable to contribute to local and regional development. Indigenous strategies seek to make places less dependent upon exogenous or external economic interest. In this case, resources are mentioned as land, natural resources, the resident local labour force, historical rooted skills, and local entrepreneurship. Contrary to Tödtling, Pike et al. (2006) do not differentiate between endogenous and indigenous approaches, but rather they use both the terms identically. Nevertheless, Tödtling (2011: 333) argues that there is a certain difference, but no clear borderline between indigenous and endogenous development concepts. However, the endogenous approach is broader than the other (as mentioned above). In term of Tödtling's perspective the development is driven in a bottom-up manner by endogenous forces and factors. In this context endogenous, but non-indigenous factors such as infrastructure investments, schools, training organizations, universities and research organizations, are purposely created or upgraded by policy makers and related institutions. As a result, highly educated workforce and knowledge and technologies are developed in the region that might drive new products, processes or other new solutions. Furthermore, endogenous forces refer to social and political factors such as involvement of social agents and civil society, which stimulate processes of self-help, local initiatives, and social movements intending for the improvement of living conditions in a particular region. Such a development strategy is called the bottom-up approach because the role of local forces and factors are strongly involved. This idea of regional development is initiated and implemented mainly by local and regional actors and agents instead of central government or external agencies, and it is intended for the needs and objectives of the regional population (Tödtling 2011: 333).

Due to actions initiated at the local and regional levels several advantages are gained compared to traditional top-down approaches such as a direct problem notion, a high intensity of interaction, regional synergies, regional strategy formulation and collective learning within regional networks (Tödtling 2011: 334).

To get a clear overview of the differences between bottom-up and traditional top-down approaches some points, outlined by Pike et al. (2006: 16–17), need to be reviewed here:

– Planning: In traditional top-down approaches the aspect 'where and how' to implement a development strategy is typically taken by national central government planners and developers with little or no involvement of local or regional actors. In contrast, in bottom-up approaches a strong involvement of local and regional actors is an integral part of the strategy. In other words, all stakeholders engaged in the region through vertical and horizontal coordination are involved. In this case local, regional, national, and supranational or international institutions are vertically coordinated and horizontal coordination comprises local public and private actors involved in development issues. Such a concept in which all stakeholders are involved can be applied everywhere, when it is carefully tailored to the circumstances of the local/regional situation.

– Implementation: While traditional top-down approaches have typically developed specific industrial sectors which in general promise to generate economic dynamism (e.g. steel, textile), bottom-up approaches adopt a territorial concept to achieve economic development, i.e. first investigate the economic, social and institutional conditions of every territory and identify the local potentials upon which proper development strategies are set up.

– Incentives: Traditional top-down approaches basically trust financial support, large incentive packages and subsidies to attract and maintain economic activities in the localities and regions. In contrast, bottom-up approaches attempt to ignore such kind of incentive packages and focus on the improvement of basic supply-side conditions for the development and attraction of future economic activity.

Stöhr (1981: 43–67), Pike et al. (2006) and Tödtling (2011: 335) also point out that the following elements and characteristics have to be taken into consideration for endogenous development:

1. Development is taken from a long-term perspective and harmonised economic, social and environmental goals have to be considered. Economic growth means not only enhancing regional production and average per capita income, but also improving the broader socio-economic situation, including the living conditions for the poor. Development should not deteriorate the environment.

2. Strategies are based on endogenous factors and potentials in particular natural resources, landscapes and tourist sites, qualified labour forces and specific skills or competences of the respective regions, instead of external and mobile ones.

3. Agriculture, craft-based industries, tourism and other services are taken into consideration in development strategies. The complementarities and interrelations between all these sectors and the strengthening of input-output linkage are formulated.

4. Development problems and growth potentials of small firms as well as entrepreneurship and new firm formation have been emphasized strongly.

5. Innovation plays an important role and is broadly defined together with technological, business and social innovations aiming at high quality of products and processes as well as the solution of broader social and other problems.

6. With respect to regional specificities in culture, local demand and economic structure, relevant contexts and factors are taken into account to achieve the unique competitive advantages for the regional firm. Such regional identity is a favourable factor for regional development by doing the branding of regional products and using new ways of marketing such as direct selling (e.g. agricultural or handicraft products).

7. Besides the economic factors, relevant social and political forces are taken into consideration. This includes activities of social agents, and the engagement of civil society supporting processes of self-help, local initiatives and social movements aiming at the improvement of living conditions in the regions. Decentralized decision making and policy competences at the local and regional levels is a favourable element for economic development, with respect to a better understanding of problems, barriers, and potentials for regional development as well as the needs and goals of the regional population.

Therefore, various spatial levels and respective institutions for each specific process play key roles for endogenous development. For example, at the local level entrepreneurial processes, economic and social initiatives and movements and at the regional level activities of regional governments and associations, official programmes for endogenous economic development, cluster initiatives and innovation support by university-industry collaboration are identified (Tödtling 2011: 334). Endogenous local and regional development is closely connected to national political and institutional structures (e.g. national economic policies), macro regional conditions and institutions (e.g. EU or Southeast Asia regional policies and structural funds) and global regimes and institutions (e.g. trade regimes and related institutions).

Tödtling (2011: 337–340) sees in particular the four following interdependent elements: industrial districts, local entrepreneurship, regional learning and regional innovation system as fundamental for an effective and promising endogenous development strategy:

1. Industrial district: In industrial districts firms are strongly embedded in the local and regional economy. This embeddedness consists of input-output linkages, but for instance also of knowledge exchange and learning as well as social relationship.

2. Local entrepreneurship: Local entrepreneurship is seen as basic for endogenous development. New firms should be originated from the region by using local talent and labour. And by doing so more local and regional input-output and knowledge links do exist.

3. Regional learning: Regional learning is a key element because it contributes to improve local competencies and skills and help to share knowledge and best practices. The result is a collective enhancement of know-how and an upgrading of practices and technology.

4. Regional innovation system: Such systems are evidently based on local knowledge and competencies and involves new products, processes and organizational practices. Such a system has to be organized well and needs elements

like knowledge transfer, innovation finance, and support for implementing the innovation.

According to Tödtling (2011: 340) two aspects of critique to the endogenous development approaches have to be pointed out:

1. It is created almost only as a counter-concept of the top-down model and lacks coherence as a theory of its own.
2. Initially it strongly focuses on endogenous factors and actors, neglecting the successful regional development that is usually the outcome of both endogenous forces and external factors such as infrastructure, mobile capital, technologies, talent and knowledge.

Therefore Tödtling (2011: 340) sees concepts of endogenous development as 'islands of development' in a world of increasing social and economic interdependencies at all spatial levels.

Keeping in mind and not neglecting the critiques pointed out above, the research is carried out based on the understanding that an effective regional development solely stressing endogenous forces is not working, but rather regional development strategies need to pay attention to both endogenous and exogenous factors and processes as well as their interaction. Such an understanding acknowledges that a diversity of development paths and development models do exist. In addition, regions normally possess a different capacity for endogenous development, and insofar they have different needs for centre or external assistance (Tödtling 2011: 336–337).

2.1.4 Sustainable development

Local and regional development activities have become increasingly important throughout the world since the 1960s and 1970s. At the same time the context has been dramatically reshaped by changes in the manner how economic activities are carried out. One of the characteristic changes is that the focus has shifted from the quantity of development to a qualitative aspect, which for instance means that economic development has to consider the necessities of the natural environment and therefore the economic development has to take into account some constraints (Stimson et al. 2009: 3–4). Furthermore, the idea of local and regional development has broadened into the direction of questions dealing with the quality of life (Pike et al. 2006: 3). This new concern is highly related to sustainable development. Insofar, sustainable development has become considerably important for concepts of local and regional development (Pike et al. 2006: 4).

According to Rodriguez et al. (2002: 3) the importance of sustainability can be defined in two ways: firstly, "[…] sustainability can be seen as a necessity in order to avoid the costs of deteriorating social, environmental and economic systems", secondly, "[…] sustainability can be seen as a source of new opportunities to improve the rate and extent of human development".

Many definitions of sustainable development do exist. They are not only based on a reflection of the complexity of defining sustainability for a wide variety of actors (from individuals to communities to organizations), but also signal a mounting concern over the deteriorating health of natural and social systems and a growing recognition of the economic benefits of sustainability (Rodriguez et al. 2002: 2). According to Rodriguez et al. (2002: 2–3) these definitions differ in scope depending on whether they are designed for individuals, companies, or national governments, however most of the definitions of sustainability are based on three fundamental premises:

1. Availability of critical input such as ecological (renewable resources), material (non-renewable resources), human (knowledge, income, health, human rights) and social factors (trust, equity, etc.),
2. Limits to the availability of finite material resources and to the regenerative capacity, or carrying capacity of ecological resources,
3. The interplay of three interdependent complex systems (ecological, social, economic), which are heterogeneous, dynamic, non-linear (more detail later and see figure 2.2).

Most sustainability definitions based on the above three premises involve two critical elements as Rodriguez et al. (2002: 3) have pointed out, namely

1. that sufficient supplies of the ecological, material, human, and social resources have to be ensured in order to allow humans to meet basic needs and to support continued development, and
2. that equitable access to this sufficient supply of resources has to be ensured both inter-generationally (among all members of the current generation) and intra-generationally (between this and future generations).

The best-known definition of this concept is from the Brundtland Commission Report, *Our Common Future,* (World Commission on Environment and Development 1987: 43), "[…] development that meets the needs of the present without compromising the ability of future generations to meet their own needs". It contains two key concepts:

"The concept of 'needs', in particular the essential needs of the world's poor, to which overriding priority should be given"; and

"The idea of limitations imposed by the state of technology and social organization on the environment's ability to meet present and future needs" (World Commission on Environment and Development 1987: 43).

Thus, the goals of economic and social development must be defined in terms of sustainability in all countries – developed or developing, market-oriented or centrally planned. Their interpretations of sustainability can vary to a certain degree, but must emphasize certain general features and must respect a consensus on the basic concept of sustainable development and on a broad strategic framework for achieving it (World Commission on Environment and Development 1987: 43).

Gibbs (2002: 3) also states that definitions of sustainable development vary, but points out the following core principles:

1. Quality of life (including and linking social, economic and environmental aspects),
2. Care for the environment,

3. Thought for the future and the precautionary principle,
4. Fairness and equity,
5. Participation and partnership.

While the discussion of the definition of sustainability continues and evidence mounts that sustainability is of critical importance for continuous human well-being, a growing number of entities – ranging from individuals to organizations, institutions, corporations and governments – have started to define their own responsibilities toward global sustainability (Rodriguez et al. 2002: 6). Particularly, these different groups articulate more precise principles of sustainability to guide their actions and make internal and external commitments to improve the sustainability of their operations. Furthermore, they also identify ways to access and report on their performance towards sustainability targets (Rodriguez et al. 2002: 6).

Although in the past sustainability has been reduced quite often to the dimension of the 'natural environment', the concept itself is more broadly structured as the foregoing discussion have shown. It is primarily focused on how human beings or a society can shape/construct its existence in such a way that it leads to stable/durable well-being. In the discourse about the concept over time, the probably most frequently used model is that of the so-called overlapping circles of the three dimensions 'Environment', 'Society' and 'Economy', which are regarded as the essential dimensions contributing to sustainable development – providing they are well arranged and interact in the right way.

Based on the model of overlapping circles (see figure 2.2) acting in a sustainable way comprises:
– Social dimension: all people shall not only have a share in societal development (like education, health care), but they also shall actively participate in designing the development, for example to have good living and working conditions, to have the opportunity of co-determining in the local community.
– Environmental dimension: working and living are organized in a way, which fit the necessity of the nature, for instance through short distances from production to supply, respecting circuits of material and using resources efficiently, reducing or preventing waste and pollution in all forms.
– Economic dimension: acting economically is organized in a way that all people can enjoy an adequate standard of living without endangering the bearing capacity of the planet, which means for instance: producing in a parsimonious, but qualitative high-grade way (eco-efficient technology), reducing poverty, enhancing equity.

The three dimensions are not independent from each other but they are intensively interrelated as can be seen in figure 2.2. The dimensions do not build only the corners of the triangle, but these corners are strongly interconnected via the triangle sides. Looking to the interrelations of always two of these dimensions (the triangle sides) at a time, the result can be described as follows:
– Social-environmental interrelation: principles are for instance environmental justice and inter-generational as well as spatially related equity (e.g. accessibility to resources). This interrelation results in a liveable world.

- Economic-social interrelation: principles are for instance creating good business ethics, respecting fair trade and human rights, producing a reasonable standard of living, which contains the idea of spatial equity. This interrelation results in a just/fair world.
- Environmental-economic interrelation: principles comprise for instance organizing production in an eco-efficient way, implementing environmental accounting as a part of a tax system. This interrelation results in a (long-term) viable/liable world.

Figure 2.2: The objectives of the sustainable development

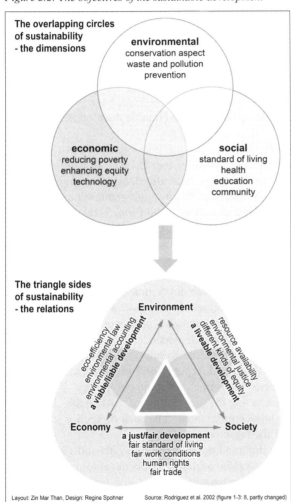

Layout: Zin Mar Than, Design: Regine Spohner   Source: Rodriguez et al. 2002 (figure 1-3: 8, partly changed)

Starting from this perspective sustainable development/sustainability will be reached when the three above mentioned interrelationships work well together, so to say the dimensions interact in a way that a situation is accomplished which is (qualitatively and quantitatively) liveable, just and fair to all people and long term viable.

This research is carried out also by keeping in mind what was mentioned by Elliott (2006: 13) namely ensuring a sustainable level of population, a critical objective, which was firstly introduced by the World Commission on Environment and Development (WCED). This aspect is obvious, because sustainable development can be expedited more easily when the population size is stabilized at a level consistent with the carrying capacity of the ecosystem. Also it has to be kept in mind that according to Elliott (2006: 29–20) two environmental philosophies do exist, the technocentric and the ecocentric one. Both are very different in the idea how sustainability can be reached. Whereas the technocentric philosophy is widely using new technique, but it is initially based on the present economic system, the ecocentric philosophy uses alternative and appropriate technologies based on a radical and fundamental change of economy and society.

2.1.5 How to combine endogenous and sustainable development

The discussion of endogenous development in the previous sections (2.1.2 and 2.1.3) has shown that the sustainability principles (see 2.1.4) like careful dealing with resources, participation of local people or durability/long-term consistency of development are closely related. In order to clarify how endogenous and sustainable development concepts can be combined, first of all a short look is taken at the older generation of policies of the 1950s to1980s and their relations to sustainability.

Vazquez-Barquero (2003: 176) clearly points out that the first generation of development policies (in the 1950s and 1960s) targeted at building infrastructure and creating attraction of external firms through incentives. Pike et al. (2006: 14) confirm this statement and argue that, if the main development bottleneck of an area is rooted in poor accessibility, heavy investment in transport and communications infrastructure can solve the accessibility problem and, as a consequence, generate internal economic dynamism and attract much needed foreign investment. But sustainable development concerns are not at the forefront of such approaches to local and regional development at this time. Storper (1997: 188–190) also mentions that according to past experiences the same development policy in different contexts has frequently no impacts on the generation of sustainable local and regional development and long-term employment. The idea of such a strategy is mainly a quick radical change of the regional economy.

The second generation of development policies (in the 1970s and 1980s) focus on initiatives encouraging the improvement of non-tangible development resources through firm incubators, business and innovation centres, and technological institutes or training centres, but they still lack the content of sustainability (Vazquez-Barquero 2003: 176).

After that the new generation of development policies/endogenous development policies has emerged. It emphasizes initiatives that prefer the creation and development of networks among firms, organizations and institutions located within the territory or in other strategically complementary territories. Its strategy really aims to improve the efficient allocation of public resources, to fairly distribute the wealth and employment and to satisfy both present and future needs of people by adequately putting natural and environmental resources to use (Vazquez-Barquero 2003: 177).

Furthermore, according to Vazquez-Barquero (2003: 177–178) the endogenous development strategies clearly differ from the first generation of traditional development policies whose main objective was to balance regional disparity existing in the territory. It was considered impossible in a global economy for local and regional administrations to adopt a development strategy based on a full spatial equity within the whole system. The new strategies expand the proposals of the second generation of policies, whose objective was to encourage entrepreneurial capacity within the territory and improve corporate productivity and competitiveness. And they aim to improve the milieu in which the productive system is added, thus converting territories into spaces for living and producing of present and future generations, according to the Brundtland report (World Commission on Environment and Development 1987). Insofar, concepts of endogenous development inherently contain elements of sustainability by definition.

How to make a productive system more competitive? The answer might involve restructuring of the productive system in such a way that agricultural, industrial, and service activities will increase their productivity and improve their competitiveness in local and external markets with the help of a step-by-step strategy that combines actions intending to the goals of efficiency, equity and ecology over the short and long term. Such a strategy coincides "with adjustments of each territory's institutional, cultural, and social system to changes in the competitive environment" (Vazquez-Barquero 2003: 144).

However, increased productivity and competitiveness can also be achieved with the help of the above mentioned radical change strategy, which just focused on the main goal: increasing competitiveness of the local productive system, no matter what the consequences are on employment, the system of organization of production, the environmental and local culture. In contrast, the step-by-step strategy encourages using the know-how and technological culture existing in the territory, and developing changes of the productive structure, which is already in existence. Such a strategy also combines the introduction of innovations with maintaining jobs and it brings about transformations in such a way that they are accepted, adopted and led by the local society (Vazquez-Barquero 2003: 144). This is why Vazquez-Barquero (2003: 145) prefers the adoption of step-by-step strategies as a best practice for both urban and rural areas. Such a strategy contributes to a viable world.

While step-by-step strategies seem to be the best practice, risks exist, too. Among others, the local economy may slide into a welfare economy model with the consequence of requiring financial support for a long time. In addition, the objectives of efficiency, equity and ecology can come into conflict with each other within the territory to achieve their respective goals. For example, the ultimate goal of efficiency is competitiveness, for ecology is the conservation aspect and equity intends to cohesion. But when one is given priority, then the other two would be constrained to reach their objectives, therefore the development strategies must reach a balance, in any case, set up priorities among objectives and actions (Vazquez-Barquero 2003: 144–177). This is why Pike et al. (2006: 18) also recommended that only the successful design and implementation of a balanced local or regional development strategy can contribute to generate social, economic and, in many cases, environmentally sustainable development.

How to adjust the institutional, cultural and social system of each region? According to Pike et al. (2006: 158 & 19), the development of policies interventions has become more sensitive to the relationships between economic, social and ecological issues. Insofar, from the perspective of longer-term outlook the development strategies should help to make local and regional institutions more transparent and accountable and foster the development of the local civil society. And these development strategies empower local societies and generate local dialogue and therefore contribute to a liveable world. By using the development strategies people, living in the area, will start to develop a degree of autonomy and adopt a more proactive stance concerning sustainable development and their own economic, social and political futures.

From an economic point of view, Vazquez-Barquero (2003: 178) states that the endogenous development approach reinforces the ecological dimension. Environmental protection becomes a source of opportunity leading to the creation of firms and jobs such as in organic agriculture, whose products are increasingly demanded and generate higher income, and in urban and rural tourism, which attracts travellers and tourists and leads to the conservation of the historical, cultural and environmental heritage. Furthermore, research and production activities based on renewable energy sources in the area are generated and service activities and technical assistance for environmental protection are created. Such a way of combining the endogenous development policy with the sustainable aspect proposes to attain multiple objectives by simultaneously promoting sustainable economic and social development and leads to a viable and just world.

Moreover, the promotion of more sustainable management of indigenous assets and resources in ways that encourage locally and regionally appropriate and sustainable forms of policy must be sought. And the differentiation between 'weak' and 'strong' sustainable development policy interventions made by Chatterton (2002: 552) and Roberts (2004: 129) should also be taken into account. According to Chatterton (2002: 559) weak sustainable development policy interventions consist of the use of environmental regulation and standards to develop new businesses, local trading networks and ecological taxes on the energy, resource use and pollution. Moreover, Chatterton (2002: 554–555) goes a step further and defines strong sustainable development policy interventions as such ones, which seek small-scale, decentralised and localised forms of social organization that promote self-reliance and mutual aid.

Certainly for regional development a way following strictly one of the philosophies mentioned in 2.1.4 cannot be implemented successfully. Moreover, both philosophies have to be combined into a compromise, which – following Gibbs (2002: 7–12) – might be called ecological modernisation. Based on the underlying social and economic system a restructuring of production and consumption is organized considering the local and regional necessities.

In sum, diffusion of knowledge, efficient productive organization, sustainable development and institutional flexibility should be stimulated simultaneously in order to make feasible the interaction among all actions/activities/strategies of endogenous development, which encourage to achieve a better competitive position of and in the territory. In this way the foundation for economic and social progress in the area is laid.

2.2 INTEGRATING THE CONCEPT OF ENDOGENOUS AND SUSTAINABLE DEVELOPMENT IN THE AREA

Regions are defined as "meso-level entities operating, in political and administrative terms, between local and national governments" (Cooke et al. 2000: 2). For instance, many regional governments of European Union member states (e.g. Germany, Austria and Belgium) can manage in a self-determined way to a certain degree the local and regional development process. This is an aim as well as a consequence of a federal system, which results in equivalent powers amongst regions, and the policies are not always designated top-down (nation-states) but bottom-up (Vazquez-Barquero 2003: 136). Endogenous regional development approaches as they have been discussed above need such self-determination to a certain degree in order to function properly.

In Myanmar the question of how local and regional development is governed has emerged as an important issue in the last years. Myanmar is a union country with a history of centralised government and limited local autonomy. Regional governments were elected recently and authorities began to gain power in 2010. But even then regional/local authorities were instruments of the central government to a great extent. In particular, the improvement of education, health, and infrastructure budgets is totally outside the control of local/regional authorities and depends on the allocation decision of the central government. However, when in 2012 the peace discussions began the government also started to create more inclusive institutions of government and to foster local democracy at the state and regional level. According to Thant Myint-U (2016) the discussion is under way and can advance in the near future successfully.

Following Assche and Hornidge (2015: 17–28), rural development can mean improving agriculture, but also moving towards multi-functional land use and diversified economy such as resource extraction, conservation, recreation, retirement, etc. In this context, the research area has to be explored in order to identify specificities and uniqueness, which can become potentials of area development or source of competitive advantages of the area. Based on the characteristics of Indawgyi Lake Area already named in subchapter 1.3, agriculture, fishery and ecotourism are specific regional sectors and the question is how they can be improved, so that they contribute to development. Keeping in mind the concepts of endogenous and sustainable development approach: the goal of all developments should be improving the well-being, reducing poverty, considering long-term perspective and respecting the ecological and environmental situation of the Indawgyi Lake Area as an Asian Heritage.

The potentials of the research area (in particular in agriculture, fishery and eco-tourism) have to be developed according to the idea of endogenous development. From this perspective the following four questions have to be tackled in creating a concrete development concepts:

1. How to initiate local entrepreneurship?: Firstly, according to Stöhr (1990: 17) societal incentives and rewards have to be offered for individual initiatives and entrepreneurship whose orientation benefits the local society. This immediately raises the question how loans can contribute to agriculture, fishery or eco-tourism development in the area. Another question is concerning the charac-teristic of the land revenue system (do farmers have land use rights and so on?). Secondly, the involvement of social agents and civil society in development programmes, which motivate processes of self-help, local initiative and entre-preneurship, needs to be organized. Thirdly, to strengthen local and regional input-output knowledge links towards learning or innovation systems cannot be ignored.

2. How to provide a regional and local learning system?: Training for human resources is necessary to enhance the local competencies and skills and share knowledge and best practices. For example, in the agriculture sector the know-ledge of farmers has to be upgraded regarding farming practices (e.g. techno-logies, products, food processing, crop-rotating systems). In the fishery sector in particular the knowledge of regulations for proper fishing techniques and fishing season needs to be enhanced, and in the eco-tourism sector an awareness of a balanced relation between supply and demand concerning environmental necessities has to be brought to the people.

3. How to create a regional and local innovation system?: Such a system is important to achieve new products, processes and organizational practices and is based on local knowledge and competencies. Obviously, knowledge transfer, innovation finance and facilities for implementation are necessary, which means that the involvement of all levels (international, national, regional and local) is required.

4. How to manage industrial districts in the area?: All firms should be rooted in the local and/or regional economy and local culture. For example, flexible labour (home work, temporary jobs) can be introduced as well as female labour. The foundation of cooperatives is another opportunity, and hiring local workers normally leads to better conflict behaviour of unions (because worker are very integrated into the local culture). All firms should also incorporate not only input-output linkages but also knowledge exchange and learning as well as social relationships.

Of course, a factor like infrastructure investment (transport, school, training orga-nization, universities, hospital, etc.) has also to be taken into account. Implementing such a factor depends strongly on the assistance of national institutions. But further-more, the main endogenous forces for implementing are related with regional and local social and political factors like participation of local people or different insti-tutions of the local and regional level. Such a strategy fulfils the idea of the bottom-up approach to a great extent.

This research is carried out following the statement of Tödtling (2011: 336) that the regional and local development is never the result of endogenous forces only, it is always the outcome of both endogenous and exogenous factors and forces, and their interaction.

2.3   KEY RESEARCH QUESTIONS

According to the above theoretical and analytical background, the empirical research aims at understanding the socio-economic development potentials via the following key research questions:
1.   Questions related to the current situation in the Indawgyi Lake Region:
     a)   How are the demographic, social, infrastructure and economic situations?
     b)   How is the governance?
     c)   How is the conservation?
2.   Questions related to evaluations by local people:
     a)   Current socio-economic situation in general, in the village and household?
     b)   Future socio-economic situation in general, in the village and household?
3.   Questions related to possible development paths:
     a)   What are future opportunities for developing the region based on its socio-economic situation?
     b)   How can local people be involved in the socio-economic development, in particular in eco-tourism?
     c)   What are the future threats of the region and which are possible ways of dealing with them.
     d)   Which are the present critical aspects in respect to the improvement in the future, and what kind of solutions can be implemented to overcome the problems?

3 CONCEPTUAL AND METHODOLOGICAL DESIGN OF THE RESEARCH

This chapter presents and justifies the research design. It describes, how the data collection has been conducted and which methods have been used in analysing the data. The chapter is subdivided into five subchapters. Subchapter 3.1 explains the research strategy, in particular it discusses the research approaches. Subchapter 3.2 outlines the research phases and subchapter 3.3 describes the data gathering and the sampling of the villages within the research area as well as the quantitative and qualitative methods used. Subchapter 3.4 presents the data analysis process and subchapter 3.5 discusses the ethical considerations related to the fieldwork of this study.

3.1 RESEARCH STRATEGY: MIXED METHOD APPROACH

The research project applies a mixed method approach, which aims at combining quantitative (statistical) analysis with qualitative background information. The reason is: the problems make different perspectives of 'looking at' necessary (Morse and Niehaus 2009: 9–15). And by using the mixed method approach a more comprehensive understanding of the situation can be reached. According to Bryman (2012: 637), a more complete answer to a set of research questions can be received better by including both quantitative and qualitative methods. Even though each approach has its own limitations, they can be compensated by using an alternative method: "The gaps left by one method can be filled by another" (Bryman 2012: 637). Thus, a combination of quantitative analysis and narrative description of the phenomenon can support a better understanding of the phenomenon in question. Such a procedure means that, to carry out the study, a mixed method design will be used here.

Critique to such a mixed method approach is connected according to Flick (2014: 36), Bryman (2012: 629 ff.) and Morse and Niehaus (2009: 41) to two inter-related aspects: "[…] 1) the idea that research methods carry epistemological commitments, and 2) the idea that quantitative and qualitative research are separate paradigms." A first conclusion then is that a mix of qualitative and quantitative research methods is not feasible or even desirable. Indeed, research methods are rooted in basic scientific paradigms like positivism (quantitative methods) and interpretivism (qualitative methods). But it is not really clear whether both method approaches can be considered as paradigms, which are incompatible and incommensurable (Flick 2014: 36). According to many scientists, in this study the methods are taken as tools to look at the phenomena in question from different perspectives. Based on this, emphasis has been placed upon the strengths and weaknesses of the qualitative and quantitative research method, then possible responses or ways in which quantitative and qualitative research can be combined, should be proposed (Bryman 2012: 629). "Such a strategy would seem to allow the various strengths to be capitalized upon and the weaknesses offset somewhat" (Bryman 2012: 628). As a result, mixed methods research becomes considered feasible and desirable. It is probably the reason that the amount of such kind of research has been increasing since the early 1980.

To address these critiques, the following research uses a triangulation to mutually corroborate the validity of research findings deriving from both quantitative and qualitative methods (Bryman 2012: 633). Facilitation includes using the quantitative finding to explain the qualitative data and/or inform about qualitative needs and vice versa (Morse and Niehaus 2009: 40–45, Bryman 2004: 457) Complementarity indicates qualitative finding that is not assessable by quantitative method and vice versa (Bryman 2012: 637, Morse and Niehaus 2009: 40–45).

Figure 3.1: The mixed method approach in this research project

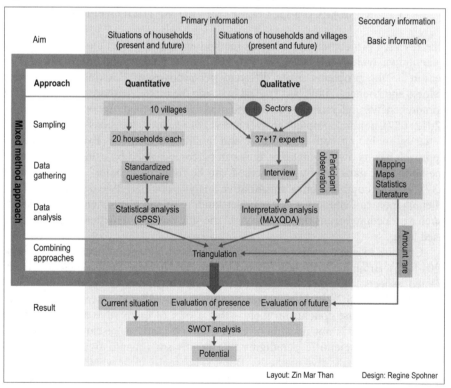

In bringing the concept of the mixed method approach into detailed and concrete action two criteria have to be considered. These are 'the priority decision' (is the main data-collection approach qualitative or quantitative or equal weight?) and the 'sequence decision' (does the qualitative method precede the quantitative one or vice versa or is it carried out parallel?) (Flick 2014: 35–37, Bryman 2012: 632). In this research the collection of quantitative data was undertaken before most of the qualitative interviews have been carried out. Yet, both data sources played an equal role in conducting the whole analysis and were combined by triangulation, facilitation and complementarity (Bryman 2004 and 2012: 632–640). The reasons are:

1. There was a need to investigate quantitatively the situation of demography, and socio-economy of the area.
2. These quantitative insights should be considered in the design of the qualitative expert interviews.
3. A follow-up research phase should be conducted in a qualitative approach to better understand the socio-economic development potentials of the area.

Overall, the aim of this study is to understand the current demographic, social and economic situation of the area, which were conducted and observed by quantitative and qualitative approaches and to investigate and discuss the development potentials of the area, which were mostly conducted by a qualitative approach, namely a SWOT analysis. A detailed overview of the whole strategy is shown in figure 3.1.

3.2   RESEARCH PHASES

The research project started in October 2013 and in a first step systematic literature search (in libraries and research publication data banks) and quantitative secondary data collection were conducted. Additionally the household questionnaire and the structure of the expert interviews were set up. Table 3.1 shows the chronological overview of the main research activities.

Table 3.1: Overview: milestones of main research activities

Time period	Research activities
October–December 2013	Literature review and preparing the first fieldwork campaign
January–May 2014	First period of fieldwork
June 2014–September 2015	Quantitative and qualitative data analysis of survey findings and interviews, writing of draft chapters and preparing for second fieldwork campaign
October–December 2015	Second period of fieldwork
January–September 2016	Continuing the qualitative and quantitative data analysis, writing of chapters, finishing of final thesis

In the first empirical data collection period both (quantitative and qualitative) approaches were carried out. From January to February 2014 the focus was laid on the systematic quantitative survey. Mainly related to the availability of experts the interviews with them were partly conducted during the time when the quantitative survey was carried out and then further continuously up to May 2014.

The second field period (from October to December 2015) finalized the field research with a qualitative focus. Having two phases of field research was very useful because it provided time to reflect on the collected data and improving/upgrading of the collected data could be carried out in a more optimal way. In particular, further data was collected to complete the qualitative research approach.

3.3   DATA GATHERING DESIGN AND PROCEEDING

3.3.1 Sampling of the villages

In order to consider the spatial heterogeneity of the lake area and to receive a deeper understanding of the variability of socio-economic situations within the area, a multi-stage sampling was employed (Bryman 2004, Flick 2014: 170–171). The first stage of sampling consisted of a non-random selection of the villages in which data collection should be carried out. The second stage of sampling included a stratified random sampling of households in the selected villages for carrying out a (quantitatively orientated) questionnaire (see section 3.3.2). The third stage of sampling was to identify interviewees within the selected villages as well as on the district, state and national level to conduct (qualitatively orientated) interviews (see section 3.3.3).

All villages are located outside of the conservation area except 60 housings in Mamomkai village (IDGY-05[7]). The non-random sampling of villages was carried out based on some specific reasons and altogether ten villages were selected for household interviews (and a great part of the expert interviews). For instance, Nyaungbin and Main Naung were selected because they developed economically differently from the other villages around the lake and – based on information of former research (Zin Mar Than 2011) – seasonal migrants were found in these villages. In addition, Main Naung is not directly located on the bank of the lake and inhabitants very rarely depended on fishing. The reason to choose Loneton village was that it could be upgraded as an administrative village for the lake area (IDGY-06). Nevertheless, in general the characteristics of the villages around the lake are more or less the same, thus the rest of the villages were chosen by the principle that the whole shoreline of the lake is covered (cf. figure 1.2). As a result, the distance between the survey villages is maximum six miles and minimum two miles. Four of the villages are located on the east side of the lake, namely Lonesant, Nammoukkam, Hepa and Shweletpan, and six villages are located west of the lake, namely Nyaungbin, Nammilaung, Nanpade, Loneton, Mamomkai and Main Naung.

---

7   The meaning of such reference is explained in section 3.4.2.

3.3.2 Quantitative approach

The quantitative approach was primarily based on a household survey. Before carrying out the survey, interviews were conducted with local community leaders to get general information about the villages. Then, a stratified random sampling of households was carried out based on the consideration of economic activities (farming, fishing, self-employment, casual work, and government or private sector staff) in the village as they have been provided by the village head. Furthermore, minimum, medium and maximum productive capacity of each economic sector was also considered in the selection procedure to understand the different levels of household income and the economic context. Based on this strategy at least 20 households have been chosen in each of the villages in order to carry out the questionnaire (cf. table 3.2).

Table 3.2: List of the quantitative interviews

Village name	Household	Household member
Nyaungbin	21	134
Nammilaung	22	130
Nanpade	22	133
Loneton	25	130
Mamomkai	21	130
Main Naung	21	123
Shweletpan	20	105
Hepa	20	112
Nammoukkam	23	133
Lonesant	21	117
Total	216	1247

216 households with altogether 1,247 household members were surveyed in the research area. That means the survey covered 6.0% of the total of 3,594 households in the ten villages and the 1,247 household members made up 6.6% of total population of the ten villages (19,014).

The household survey was set up based on ideas of Abdrabo and Hassaan (2003), Bhat (2005), Cinner (2000) and Szirmai (2005) with the aim to deliver data on three main issues (see Appendix 1): the first part of the questionnaire (questions 1 to 30) was related to the current demographic, social and economic situation of the households and household members. The second part of the survey included question number 31, which dealt with the evaluation of the current socio-economic situation. The final part of the survey consisted of a question complex (number 32), which was related to four topics: firstly, ideas of the households for a future economic plan; secondly, the assessment of future socio-economic development of the households and of the village; thirdly, their interest and involvement in eco-tourism and fourthly, main critical aspects deserving improvement for area development.

In more detail the questionnaire was structured as follows (see Appendix 1):

Part 1: Data collection about the demographic, social and economic situation in the research region: This part of the questionnaire is similar to a census (see 3.3.4 regarding the census situation in Myanmar), in which information about the general situation of the region is recorded. The collection of information on the region through the households was necessary, because at that time almost no data for this region and only limited information about the present socio-economic situation existed. The following data was collected:

– Demographic factors: family size; ethnicity and culture; gender distribution; age, and their personal data (occupation and place of birth); living condition, in respect with homestead types, electricity-, water-, and information sources, access to communication, ownership of transport devices, entertainment programme, and waste management practice, and migration in respect to residence time, reasons and pathways of migration.
– Social factors: health situation regarding access to health care facilities; current individual health problems in the household; awareness of health (e.g. treatment behaviour, sleeping with mosquito net or not, etc.); educational situation in respect to individual educational status, and access to education facilities.
– Economic factors: main economic activities of the household; annual expenditures of the household; income sources of the household and land/livestock/farming machine ownership.
– Infrastructure: availability of transportation; communication and information; electricity supply; water supply; health facility and education facility.
– Firewood consumption and fishing gear in use are also recorded to identify impacts on the natural resources. These two mentioned aspects are also strongly related to economic aspects like expenditure and income.

Part 2: Evaluation of the current socio-economic situation by the local residents: This part contributed to make the situation described by the above data more tangible. It provided indications of the positive and negative developments as seen by the local people. Additionally, it provided evidence of the region's strengths and weaknesses. The evaluation measurement is classified into four levels ('good', 'tending to good', 'tending to bad' and 'bad'). The signal items were set up to understand the evaluation of the current socio-economic situation (including income generating, job opportunity, market accessibility, water quality, health facility, education facility, impact of conservation area on local economic and current eco-tourism) as well as the current infrastructure situation (including transportation, communication, information facilities, electricity system, farm water availability and waste management).

Part 3: Assessment of the future: This part went a step further and presented, how the future was seen by local people and highlighted also the interest of the households in eco-tourism as one of the economic activities. Altogether, in particular four aspects were of interest:
1. ideas of changing the economic activities of the household in the future,
2. assessing the future situation of the own household as well as that of the village,

3. future interest of the household in participating in eco-tourism, and
4. main critical aspects deserving improvement for area development.

All the assessments were measured again in four levels ('very positive', 'positive', 'negative' and 'very negative').

The surveys were conducted personally either during family visits or at work places (school, health centre, office, farm etc.) and some questions were answered by observation from the interviewers, e.g. housing and living conditions. Each of the household surveys took about 45 minutes. For each of the 1,247 household members, personal data e.g. education level, age were collected.

This quantitative survey was carried out from January to February 2014 with the help of one research assistant, who is a native from Mamomkai village and was a first-year distance education student of Philosophy at Myitkyina University (Myitkyina is the capital of Kachin State).

Before carrying out the survey, the research assistant was trained to understand the research objectives, to be familiar with the survey questions and to improve the interview skills. So, every survey question and all principles of conducting a sample survey were explained. Furthermore, the questionnaire was prepared in Myanmar language because the research assistant preferred it. To ensure the quality of work and to build up the research capacity of the research assistant, reflection meetings were carried out on a daily basis. In these meetings discussions about the progress, complications, weaknesses and strengths of conducting the survey took place. In addition, the research assistant was asked to add reflection notes to each interview, if necessary. These reflection notes include any relevant additional information to the main research questions.

Overall, the collaboration with the research assistant resulted in a lot of advantages. Firstly, the participation of the research assistant speeded up the process of carrying out the survey and enabled the collection of required data in a short period of time. Secondly, the research assistant was not only an enumerator, but also a resource for insight about life in the area. Furthermore, her reflection and opinion were enriching and very useful for the further data analyses.

3.3.3 Qualitative approach

The qualitative approach comprised of the expert interviews, field observations, participant observations and field notes. Sampling of experts was based on their professions and positions. Partly the experts were from the government sector and partly from the non-government/private sector. To consider different perspectives and strategies regarding the area development, experts from a variety of different sectors and from different levels (national, state, district and local) were interviewed.

Expert interviews

In the first field period interviews were conducted mainly on the local and township/district levels. Besides getting information on the situation and the problems of the area from different perspectives, the information received in these interviews helped to create development ideas, which matched the local needs and were related to bottom-up development approaches. Altogether 37 experts were interviewed:

– Government sector: administration (e.g. district officer, township officer, local authority), conservation (e.g. warden and ranger of Indawgyi Wild Life Sanctuary), education (principal of Mohnyin Degree College, Professor, heads of schools), health (physician, health assistant, midwife), fishery (assistant fishery officer).
– Non-government sector: NGOs (Deutsche Welthungerhilfe, Fauna Flora International, and Friends of Wildlife, which were familiar with the area, women affairs member and youth association member), farmers, village elders, and medical doctor (cf. table 3.3).

During the second field period 17 expert interviews (the affiliation of the experts to different fields can be seen in table 3.3) were conducted and discussions were mainly focused on three aspects:

– to understand the regional development policy/strategy of the regional and national level,
– to have discussions with local experts a) on the preliminary results from the first fieldwork, b) on development paths based on the region's potentials as well as on principles of endogenous and sustainable development and
– to upgrade/complete the findings from the first field period.

Table 3.3: List of expert interviews for first and second field period

Institution	1st field trip frequency	2nd field trip frequency
Administration	12	1
Health	7	1
Education	5	3
Conservation	5	4
Agriculture		3
Fishery	1	2
Tourism		1
Village elder	4	
NGO	3	2
Total	37	17

The construction of qualitative expert interviews was conceptually based on ideas of Morse and Niehaus (2009: 85–115), Bernard (2006: 210–250 and 2013:180–210) and Alvesson and Kärreman (2011: 98–110). For framing the concrete questions the study of Kraas (2014) on "The Myanmar Urban Network System: 81+

cities" was a guideline. Additionally, ideas of the SWOT concept[8] (Dealtry 1994) were used in order to figure out today s̲trengths and w̲eaknesses at the time of the interviews, and future o̲pportunities and t̲hreats of specific aspects (e.g. administration, demographic, migration, transportation, communication, electricity supply, water supply, waste disposal management, education, health, agriculture, fishery, gold mining, forestry, industry, handicraft production, services, tourism and recreation). The semi-structured and open-ended questions encouraged interviewees to give their opinion on various issues. The questions were always mainly focused on topics related to their relevant professions in order to gain a deeper understanding on their everyday situations and problems (see Appendix 2).

Almost all expert interviews were conducted in the Myanmar language. The interviews were recorded and transcribed. Each interview took 25 to 45 minutes. During the first fieldwork campaign, expert interviews mostly (local and district level) were done in Indawgyi area and Mohnyin. In the time of the second field campaign most of the interviews (regional and national level) were conducted in Myitkyina.

Field Observation and Participant Observation

Observation is another tool, which has been used in this research. This method is in particular elaborated in Anthropology and Ethnography. Observation includes particularly all senses – seeing, hearing, feeling, and smelling (Flick 2014: 308). Applying this method requires living together with the people being studied, experiencing the same living conditions, participating in economic and social life events of the people. Thereby the researcher can gain insights into people's reality, perceptions and problems that they face in their daily life (Chambers 1983: 75–101). In addition, useful information can be obtained through relaxed informal conversation and observation.

While living in the villages during the field campaigns two methods of observations were employed in the present research. The first was observation of subjects without engagement in their activities (especially their economic activities: fishing, farming, mining, etc.). The second was observing through direct participation in the activities (e.g. making snack, harvesting peanut). Participant observations were particularly useful to building trust and rapport as well as gaining proximity to the subjects (Bernard 2013: 310–311).

"Participant observation is a strategic method that lets you learn what you want to learn and apply all the data collection methods that you may want to apply" (Bernard 2013: 327). Thus, "the strength of participant observation is that you, as a researcher, become the instrument for data collection and analysis through your own experience" (Bernard 2006: 359–360, 2013: 321). While living in the village an introduction of the researcher took place simply, honestly, briefly and consistently. It enabled the researcher to get close to people and making them feel

---

8   The SWOT concept will be discussed in more detail in 3.4.3.

comfortable enough with the researcher's presence. The village life and a number of social events (e.g. celebrations for wedding, funeral, celebration for childbirth, a naming ceremony, annual pagoda festival, etc.) were observed.

Particularly, being from the same country – thus exposed to the same social culture context and the same language – it was easy to become an insider, to build friendship and to understand interviewees' daily life. However, as an insider it was harder to recognize cultural patterns that were used every day and likely a lot of things were taken for granted. On the contrary, an outsider might pick up such aspect right away and would see it more clearly (Bernard 2013: 330).

Overall, living in the village was very useful for understanding the village context, village structure (it is not only an administrative unit, but also a community), resource usage, social relations among the villages, among the people within the community as well as within households, and the awareness of environment among the household members. To gain knowledge on the aspects it is essential not only to understand/learn people's personal context (how people discussed particular issues, how they explained their reality and how they defined the meaning of their lives), but also to understand the ways people spoke: for instance which words they used, and the tone in which they spoke. In some cases, talking about sensitive issues required a special effort to encourage people to share and to engage in open communication and to feel comfortable. An acceptable way to talk about sensitive issues – especially, (illegal) gold mining and timber production as well as fishing, which ignored the closed season – were elaborated.

3.3.4 Secondary sources

Besides the quantitative approach (household survey) and the qualitative one (the expert interviews and participatory field observation) as basic information sources, some information on specific aspects was gathered from secondary data. This data included information about flora and fauna, water quality data, soil sample data and other secondary data. Secondary data were collected through literature reviews and from statistical sources. However, the interviewed employees from township administration department said that there was very few general administrative data of the area and not all of the data were reliable because the last population census was conducted long ago and the new one would be carried out in March 2014 (IDGY-07). Unfortunately, this census data could not be used in this study as a main source of information because only general results have been published, but not detailed information about the research area. Additionally, and even more important the census covers much less aspects than have been considered in the questionnaire.

3.4   DATA ANALYSIS

As mentioned above this research employed a mixed method approach to analyse quantitative and qualitative data through triangulation, which represents just one way in which it may be useful to think about the integration of these two research strategies (Bryman 2012: 392) (see subchapter 3.1).

3.4.1 Quantitative data analysis

The quantitative survey results were categorized and analysed to determine and to reflect according to the following foci:
– Understanding the current demographic, social, economic and ecological differences between individuals, households and locations,
– Understanding how local people evaluate the current socio-economic situation,
– Gaining insights about the household's future plan and how the future socio-economic development is assessed at the household and village level,
– Exploring locals' interest in eco-tourism as an economic source and investigating their possible involvement in the sector, and
– Understanding the main present critical aspects that need to be improved in the future.

The quantitative data were analysed using the statistical software package SPSS (version 22 and 23). The initial variables in the database were set up based on the semi-structured questionnaire and new variables were created based on the original ones during data analysis. Thereafter, descriptive statistics were applied to identify characteristics of specific aspects (such as age, education etc.), common tendencies and the variability among the different groups in the population/villages. The descriptive statistics primarily included the generation of frequencies and of univariate statistical parameters like mean, maximum, minimum, range, percentages and quantiles.

In particular for the analysis of the financial situation of the households Box-Whisker-Plots – or shorter Boxplot – were used to present the results. The boxplot is a graphical method, which is used to describe the distribution of a variable. The basic form of the boxplot is using the parameters of the minimum, the maximum and the quartiles (1st quartile, 2nd quartile = median, 3rd quartile). A more detailed description of the method can be found in De Lange and Nipper (2016). The result is a diagram like it is shown in figure 3.2.

The area between the first and the third quartile is established as a box and the median (2nd quartile = 50% of the values are below and 50% are above the median) is drawn as a line within the box. The median separates the box into a lower and an upper box part. The distance between both the quartiles (= height of the box) is called the interquartile range. The area between the maximum and the 3rd quartile and between the minimum and the 1st quartile are drawn as lines, the so-called whiskers. In each of the areas (lower and upper whisker, lower and upper box part) 25% of the values are located. The box contains 50% of the values. The length of both the whiskers and the heights of the upper and lower box part give information about, whether the distribution of the variable values is fairly even or whether a concentration does exist somewhere. If the lengths and the heights are similar the distribution (over the quarters) are fairly equal. If some lengths or heights are shorter, the distance between the values in this area is shorter. So, looking to the lengths and heights gives information about the variation of the values.

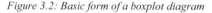

Figure 3.2: Basic form of a boxplot diagram

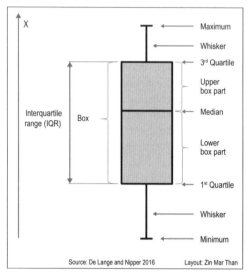

An extension of this basic form of boxplot can be done insofar, that extreme values are specifically pointed out. Figure 3.3 shows this extended version of a boxplot as it is also used by SPSS. In this version, the endings of the whiskers are not necessarily chosen as minimum and maximum, but as a multiple of the height of the box, namely 1.5 times the height of the box. The reason for this is grounded in the characteristics of the normal distribution. For such a distribution 95.7% of the values are located within the area between the endpoints of the whiskers (1.5 times the height of the box). The endpoints of such whiskers are also called upper and lower fence. The few values (about 4.3% of all values), which are not located within the fence are called outliers. In case the minimum is smaller than the endpoint of the lower whisker or the maximum is bigger than the endpoint of the lower whisker (no outliers do exist), the endpoints are set at the minimum or maximum respectively. The outliers themselves will be subdivided into such ones which are located in the interval from 1.5 times to three times (○), and in such which are located outside of the three times marking point (*). Such values are also called extreme outliers.

Such a boxplot provides a good comprehensive overview about the distribution of the values of a variable. In particular, boxplots can be used quite well for comparing variables by putting the boxplots of the variables, which have to be compared, side-by-side. Important characteristics of their distribution can be made clear graphically. In this form the method is used in this analysis.

Figure 3.3: Extended form of the boxplot diagram

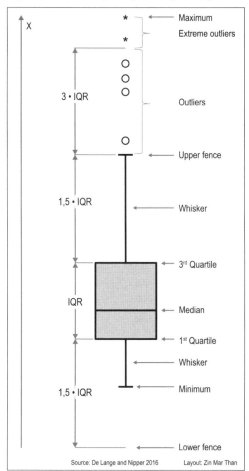

Bivariate analyses (crosstabs/contingency tables) were carried out to identify relations/correlations between different variables. For instance, this technique was useful to learn the relationships between education and migrant/gender, and between occupation and income, etc. The contingency coefficients were tested by the chi-square test. As the residuals of the contingency tables are following a normal distribution, the significance levels for them are defined in the following way:

> 2.576: the residual is significant on the 1%-level (= strongly significant)

    → the observed frequency is extremely disproportionally high

2.576 to > 1.96: the residual is significant on the 5%-level (= significant)

    → the observed frequency is disproportionally high

1.96 to -1.96: the residual is not significant

    → the observed frequency do not differ significantly from the expected value given an equal distribution

< -1.96 to -2.576: the residual is significant on the 5%-level (= significant)
  → the observed frequency is disproportionally low
< -2.576: the residual is significant on the 1%-level (= strongly significant)
  → the observed frequency is extremely disproportionally low.

3.4.2 Qualitative data analysis

The qualitative data analysis process started in the field and included several stages. First, the data were documented as field notes, recording files, narratives and observations. The interviews were transcribed (plus translated into English) and entered into the software programme MAXQDA.

Thereafter, the data was organized, coded and reflected according to the following foci:
- Understanding better what were the current weaknesses and strengths
  - of each economic/income generating sector (e.g. farming, fishing, gold-mining, casual work, services, tourism),
  - of the social situation (e.g. culture, health, education, migration),
  - of infrastructure (e.g. communication, education, health, water supply electricity supply, transport),
  - of nature/environment and,
  - of institutions.
- Gaining insights about the future opportunities and threats of the aspects mentioned in the first point of this list.
- Exploring the present critical aspects deserving improvement in the future with respect to economic situation, social situation, infrastructure, environment, and institution and investigating solutions that can be implemented to overcome these problems.

To explore a specific aspect (e.g. education), a specific code was retrieved as a single output document and analysed with the help of MAXQDA. All interviews received a code number according to IDGY-XX, where XX is a number (see Appendix 3). Direct or indirect quotes and information from the interviews are cited in the text via this code system (e.g. IDGY-23).

3.4.3 SWOT analysis

In order to evaluate the data with respect to possible development paths the results are included into a SWOT analysis. Actually, SWOT – acronym of **S**trengths, **W**eaknesses, **O**pportunities and **T**hreats – is an analytical tool for strategic management planning devised in the 1960s by Albert S. Humphrey, a business consultant at the Stanford Research Institute (Paraskevas 2013). It is one of the most popular tools for the economic sector and also a useful tool for risk management (Paraskevas 2013).

In general, SWOT is a technique to structure the current situation of a company or a field/business sector in terms of internal and external critical elements and to identify the potential areas as well as the priorities for strategic planning, in order to create a common vision of achieving the development strategy (Enache and Carjila 2009). Internal and external elements mean that researchers in the strategic management classify strengths and weaknesses as internal elements of the firm/organization like good customer services, high quality products or weak financial resources. External elements are opportunities and threats, which are not located inside the firm, but belong to the 'environment' of the firm like government regulation or the development of the new technologies (Helms 2013).

Strengths represent – according to Enache and Carjila (2009) – the favourable items of the analysed field, which have been assessed as a certain advantage to the field. Weaknesses describe unfavourable characteristics of the field/sector examined, which means lower performance compared to other fields or lack of resources it needs but does not possess. Opportunities are positive external factors, which offered to the field/sector to establish a new strategy or to reconsider the existing strategy in order to exploit profitably opportunities arisen. They exist for each area and must be identified to determine on time the appropriate strategy for their recovery. Threats are obstacles arising from the external evolution and are negative factors, which have to be eliminated and minimised.

In contrast to the original distinction made with the terms 'internal' and 'external', in this study the distinction between both the terms is made according to the temporal dimension. Strengths and Weakness as internal elements refer to the positive and negative outcomes of the current situation of the region. Opportunities and Threats, however, refer to the conditions and results, which will or might take place in the future.

Dealtry (1994) comments that some of the SWOT issues are a muddle of observation: some are taken from a purely personal point of view and some from an organization point of view. He advises to take a broader perspective and concentrate on important issues, because sometime the experts do not focus on or in some cases fully grasp the main issues that affect the future development. Also some issues, which are included, are not sufficiently important, i.e. the expert puts in what he is most familiar with, rather than what he now realizes is actually needed. To mitigate such situations Dealtry suggests that 'looking behind' will put more effort into uncovering really important issues. In addition, there are some issues which are listed that are not within expert's ability to control, for example consequence of drug issue, mercury contamination, deforestation: all may reflect on SWOT at some time in the future. But if the expert cannot influence these issues how can these issues be treated in the SWOT analysis? To solve these problems Dealtry states that they would be allocated into their respective time frames, where a strategy is being formulated, perhaps based upon different perceptions of experts. He also mentions the so-called 'back-to-back' effect: with time it is possible that a threat can be turned into an opportunity and vice versa.

Considering the above pros and cons the concept of SWOT analysis can contribute to better understand the current social, economic, infrastructure and governance situation of the study area, with the aim to promote the regional development potentials, more well based and much more determined. With respect to this idea the study (see figure 3.4) firstly appraises current strengths and weaknesses, secondly future opportunities and threats, in order to combine finally the effects of strengths, weaknesses, opportunities and threats. Based on that, potentials are identified and to get development paths the formulation of strategies is set up by applying the principles of:

– building on strengths,
– eliminating weaknesses,
– using opportunities, and
– mitigating threats.

Simultaneously the ideas of the concepts of endogenous development and sustainable development as well (cf. figure 3.4) have to be considered.

Figure 3.4: Applying the SWOT analysis in this study

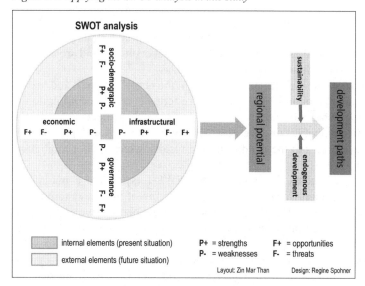

3.5   ETHICAL CONSIDERATIONS

The challenges of conducting social science research in the area were threefold: Firstly, the interviewees were not used to answer questions in the context of an interview (a few interviewees were nervous). Secondly, quite a number of interviewees were unfamiliar to interviews being recorded. These circumstances resulted partly in reserved answers. Thirdly, in some interviews the respondents hesitated to explain their financial situation (e.g. income).

Moreover a regionally unique challenge is rooted in historic political and ethnic circumstances. It has led to long-term ethnic conflicts with the effect that the trust among people of different ethnicities has been reduced immensely. Therefore, some suspect/curious questions were risen: Who sent you? Why do you want to learn about people here? What is your research good for and who will benefit from the research? Another reason for this scepticism towards the researcher could be that it was extremely rare to see researchers in the villages.

Ensuring informed consent and explaining the purpose of the research were the main ethical principles for conducting the research, which helped to overcome the above mentioned issues (Flick 2014: 49). In particular, the following steps were taken:

– introducing the interviewer as a person,
– introducing the purpose of the research (PhD on socio-economic development potentials in this area),
– mentioning that the DAAD (Deutscher Akademischer Austauschdienst/ German Academic Exchange Service) is supporting the research, and
– mentioning that all required permissions to carry out this research were received from university, district, township and village authorities.

In accordance with Flick (2014: 50), the author is convinced that adhering to ethical principles should be the top priority of any research in order to avoid causing any harm to respondents. Therefore, the interviewees were explained that they were not obligated to answer questions and they had the right to stop/conclude the interview at any time. In addition, the researcher made sure that the interviewees did not mean that their answers were ultimately recorded and written down. Any data that the interviewees did not want to be recorded/written down were not included. All of the interviewees were given the researcher's contact information in case they wanted to withdraw their participation or some data afterwards. Finally, the interviews were always conducted at that time, which was most convenient to the interviewees and the interviews were carried out with respect to the interviewee's values, traditions and culture according to the ideas of Chambers (1983: 191–198).

4   DEMOGRAPHIC AND SOCIAL SITUATION

This chapter describes results and findings of the demographic and social situation of the area. Firstly, an investigation of the basic demographic patterns (4.1) like ethnicity and age structure is done. The living conditions (4.2) are the second topic of this chapter, before the education situation (4.3) is looked at in detail. The situation of health is explained in subchapter 4.4 before the chapter ends with dealing with migration patterns (4.5).

4.1   BASIC DEMOGRAPHIC PATTERN

Population

About 74% of the population belong to the Shan-Ni ethnic group (a Shan subgroup), 20% are Burmese, 6% Kachins and the rest Chin and Kayin. According to experts (IDGY-08 and 50), most of the Shan-Ni and Burmese are Buddhists, and most Kachin are Christian. The groups live harmoniously with each other and have neither religious nor ethnic conflicts. Intermarriage, for instance between the Kachin and the Shan, is not uncommon and as one expert (IDGY-08) commented, this attitude proves a tolerant interrelationship among the ethnic groups.

Most of the Kachin speak their own language at home and children have the opportunity to learn their language in church during the summer holiday. However, among the Shan-Ni, almost all under 50 years of age cannot speak their mother tongue. In Nyaungbin, Nammilaung and Nanpade, village people rarely communicate in their own language except for some of those above the age of 50.

One respondent explained her problem in her childhood: she spoke Shan language before she started school where Myanmar language was taught, then she faced the language obstacle and was shy in the class, so gave up to speak the Shan language at home, thus she was not used to Shan language anymore (IDGY-27). However, as one expert (IDGY-24) pointed out since 2014 opportunities exist to learn the Shan language in the villages and children can learn their own ethnic language at school. Another expert (IDGY-25) commented that it is necessary to develop further ethnic cultures in the area, as all minorities feel that they have the same rights, which will lead to ethnic and national unity.

Table 4.1 gives some information on the household size based on the household questionnaires. It shows an average household size of approximately 5.8 members per household. Between the villages quite a variation exists ranging from a low average in Shweletpan (5.3 person/family) to a household size in Nyaungbin, which accounts for about one person per household more (average equals 6.4). 80.9% of the surveyed population[9] are lake natives, only 10% of the population are born in middle Myanmar, and the rest was from other parts of Kachin State.

Table 4.1: Information on household size

Village name	Number of households	Household members	Minimum	Maximum	Mean
Nyaungbin	21	134	3	11	6.4
Nammilaung	22	130	4	12	5.9
Nanpade	22	133	2	12	6.1
Loneton	25	130	1	10	5.3
Mamomkai	21	130	3	12	6.2
Main Naung	21	123	2	13	5.9
Shweletpan	20	105	2	10	5.3
Hepa	20	112	3	12	5.6
Nammoukkam	23	133	2	9	5.8
Lonesant	21	117	2	11	5.6
Total	216	1247	1	13	5.8

Figure 4.1: Age group and gender

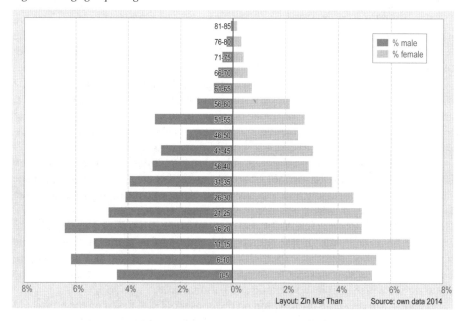

Layout: Zin Mar Than   Source: own data 2014

As can be seen from the age pyramid diagram (based on five years age groups in figure 4.1) the age structure still shows a typical pyramid form. The young age

---

9   Surveyed population covers all household members of the interviewed households: altogether 1247.

classes from 0–5 to 21–25 are the largest groups and the gender is not completely balanced in this range. The middle age classes from 26–30 to 46–50 and the old age classes from 51–55 to 81–85 show very much a constant slightly higher number of females. It is worth noticing that there is a decrease in the very young age group (0–5 years) compared to the other young age groups. It has to be investigated further whether this decrease is a first indicator of a future trend of decreasing birth rates. According to experts (IDGY-22), since 2011 there has been a successful family planning programme because of sufficient medicine and health education programmes.

According to the household survey, females slightly exceeded the males (637 females and 610 males) in the study area. Gender equality was imbalanced with respect to education, occupation, income, social activity and religion. There was a distinct difference in school level attendance with a very disproportionately higher number of females with university education, while a converse situation can be stated for the numbers of male respondents with university education (cf. table 4.2).

Table 4.2: The relation between school levels attained and gender

Gender	Monastic/ primary	School level attained		
		Middle school	High school	University
Male		++		----
Female		--		++++

Legend:
```
----    = very disproportionately low (=significant at the 1% level)
--      = disproportionately low (= significant at the 5% level)
++++    = very disproportionately high (= significant at the 1% level)
++      = disproportionately high (= significant at the 5% level)
```

However, males were dominating other sectors of the society. For instance, a respondent says

> "[…] Male can be a head of the village and I have never heard and seen that the head of the village is female" (IDGY-27).

No doubt, Myanmar never had a female country head, moreover, before 2015 only very few females were members of the parliament (less than 5% (The Gender Equality Network, 2012)). Nevertheless, today, a progress can be seen with 13.3% of parliamentary seats occupied by women due to the outcome of November 2015 election (Inter-Parliamentary Union 2016).

According to experts (IDGY-23 and 27) the area offers mainly physical work in gold mining, fishery and agriculture. So the income difference between the genders are quite high: for instance while a male gold mining worker earns K 5000 (€ 3.9[10]) per day, a female is paid K 4000 per day (€ 3.1). A respondent (IDGY-14) says that woman are very motivated to participate in the community work. On the

10   The exchange rate between Kyat and Euro is set to 1 € = 1300 K.

other hand, there is still a restriction for woman regarding religion; for example some holy places at the pagoda are only opened to men (IDGY-06).

Working status

The numbers of the occupational and non-occupational population, counted among the family members, resulted in a ratio 50.5 to 49.5: 50.5% who had occupations such as farming, casual work, other self-employment etc. (cf. table 4.3) versus 49.5% who represented the non-occupation group which consists of students and dependents, who in most cases are either too young or unable to work. Gold mining or agricultural workers, who own neither gold mine nor farmland, were very predominant in the group of casual workers. In general, other self-employment consisted of persons, who had either some investments or a capability such as retail-shopkeeper, trader, gold miner, livestock breeding or were offering services such as rice mill, boat/car/elephant rental, restaurant, carpenter, tailor, photographer, beauty parlour, workshop and so on. The occupation group 'other' consisted of people who have a job with regular income but do not depend on benefits like retirees or by contrast of people with irregular income but depend on performances such as broker, driver and so on.

Table 4.3: Occupation of household members

Occupation	Absolute frequency	Percentage frequency
Farming	245	19.6
Casual work	104	8.3
Other self-employment	102	8.2
Fishing	97	7.8
Government or private staff	63	5.1
Other	19	1.5
Total	630	50.5
Non-occupation		
Student	385	30.9
Dependent	232	18.6
Total	617	49.5

Table 4.4 shows for the adult population the relation between occupation and age. Age is categorized into three age classes: 16 to 35 (considered as young people in working age), 36 to 60 (considered as older people in working age), and 61 and above (elderly population). The contingency table proves that the age groups are significantly different in relation to the kind of occupation, the contingency coefficient is with $C = 0.4906$ quite high. Naturally, the elderly people are very disproportionally often counted as dependent as they are retired. Interestingly, the people of the two groups in working age show fairly different patterns. It might be not a

surprise that people of the younger age group are very often students, whereas, on the contrary, people from the older age group belong very often to the non-occupation group. But in general, both the groups in working age are more or less differently structured in relation to the occupational categories. The younger age group is significantly often engaged in casual work and as government or private staff. The older age group is more involved in farming and in other self-employment. Only for the occupation fishing both the groups in working age do not show a difference. Here the number of people engaged is very similar to the 'average' number, which is expected.

Table 4.4: The relation between age group and occupation/non-occupation

Occupation	Age group		
	16–35	**36–60**	**≥ 61**
Farming	----	++++	
Fishing			
Other self-employment	----	++++	--
Government/private staff	++		--
Casual work	++++	--	
Non-occupation			
Student	++++	----	--
Dependent	--		++++

Legend see table 4.2

4.2   LIVING CONDITION

The structure of the houses varied between the villages, in terms of the roof material and shape, the construction and material used for walls and floors. In particular, the villages of Shweletpan and Nanpade showed differences compared to others. The predominant roofing material in the research area was iron sheet with 72.6%, although thatch with 25.6% is quite often. The latter was used more frequently in Shweletpan and Nanpade. Wooden walls and wooden floors were predominant with 47.7% and 74.0% respectively, compared to bamboo walls with 46.0% and bamboo floors with 21.4%. Again, the bamboo construction type was more frequently observed in Shweletpan and Nanpade (cf. table 4.5).

The kitchen was mostly attached to the house with 49.3% and 34.0% built separately from the house. In some villages – again in particular Shweletpan and Nanpade – it was more frequently arranged as a makeshift outside the house with over 16%. Mostly the bathing place was arranged as a makeshift outside near the water pump with 34.9%. It was also built inside the house with 31.2 % and outside the house with 10.2%. The rest did not have a bathing place at home and took the bath either in the lake or at a nearby tube well. The predominant toilet structure used in the research area was covered pit with 98.6% and the rest did not build own latrine and shared with neighbour. The solid waste disposal practice was mainly burning the waste in the yard with 92.1% and some households (with 5.1%) deposited their wastes at the edge of the village. The rest used a pit or landfill somewhere outside the villages.

Table 4.5: Household material roof, wall and floor

Material	Percentage frequency						
	Iron sheet	Wood	Bamboo	Thatch	Brick	Other	Total
Roof	72.6	-	1.4	25.6	0.5	-	100
Wall	-	47.7	46.0	-	5.6	0.5	100
Floor	-	74.0	21.4	-	4.7	-	100

While 29.6% of interviewed households depended on solar energy for house lighting, 46.3% used electricity produced by generator (new development for electricity supply since 2015 are described in 5.3). However, candlelight was still frequently in use with 24.1%. Firewood was the main fuel source for cooking with 96.3% of total household while it was extremely rare to find households, which used electricity to cook and only very few households with 3.3% used charcoal for cooking.

Almost all households, i.e. 93.5% of interviewed households, took tube well/pump water for drinking, 6% depended on lake water and only one household, which belongs to a medical doctor, bought drinking water from Mohnyin, 33 miles away. In general, the distance of access to drinking water of the households was a maximum of 0.4 miles and a minimum of none because tube well/water pump existed in the yard of the house with 77.3% of interviewed households. 89.4% of interviewed households used tube well/pump water for household water, the rest depended on lake for household water. The distance of access to household water was more or less similar to the distance of access to drinking water source and 78.2% of interviewed households had access to household water at home. In conclusion, most households depend on tube well/pump water for both purposes: drinking and household use. Moreover, most of the households had tube well/pump at home, nevertheless the result showed that households more frequently used the tube well/pump for drinking than lake water and it was very rare to find a household, which uses separate sources for drinking and household water.

The main effective information sources of households were electronic media such as television, radio and mobile phone. 66.7% of the households used such devices. The vast majority, which did not have access to electronic media, was informed by word of mouth (= 29.1% of all households). The rest depended on print media, which were rather delayed due to transportation bottlenecks. The predominant communication system was either personal contact or via phone, 46.8% of interviewed households had access to mobile phone. 58.4% of them had one access, and the rest owned two or more per household.

Mostly the households possessed mobile phones since 2013, although the CDMA[11] mobiles have been established already in 2008 (new development for mobile phone service since 2014 is described in 5.2). Households, which did not have access, depended on the use of private telephones. Usually the owners collected charges for use. The communication via postal service, which is only located in Loneton village, was rare and just only for offices/public issue.

Regarding transport mode within the area the predominant one was motorbike. 73.6% of the interviewed households possessed at least one motorbike and 20% of them owned at least two motorbikes, while it was very difficult to find a household, which possessed an automobile; only 16 cars belonged to 12 households, i.e. 5.6% of interviewed households owned such a motor vehicle. Moreover, 27.3% of interviewed households owned some bicycles, which were in use especially by students. Besides that, there was also water transport option by motorboat.

The majority of households enjoyed watching movies in their daily life, 72.7% of the households had access to television and 70.8% had access to DVD player. 53.2% owned radio and only very few households with 10.6% possessed cassette recorder. It was also extremely rare to find a household, which owned a computer. Only five households had a computer, bought since 2010. That represented 2.3% of interviewed households. Nevertheless, since 1996 the number of entertainment equipment increased constantly over the years. The interesting thing is that mostly this equipment were bought in 2013, when the economic situation of households seemed to have improved. On the other hand, also some problems were involved because households bought such new appliances although they did not have sufficient money as one expert mentions,

> "[…] Some locals have lower income and cannot manage the income, spend lot for entertainment, when they have money, they buy television, DVD player etc." (IDGY-04).

The famous event in the area is an annual pagoda festival near Nanpade village, which takes place between the end of February and beginning of March for seven to ten days. During the festival most of the pilgrims and all shopkeepers, who come from other parts of Myanmar, camp on the bank of the lake near the pagoda, which is located on the lake's open water at the middle of its western edge. The business activity during the festival is an important opportunity for the local people to look for goods, they normally cannot buy in the area easily. So, they shop for household materials, bedding, clothing and electronic equipment during the festival, which also offers some entertainment programmes, which are rarely available for them. According to experts (IDGY-24 and 20), the villages around the lake collaborate and organize the pagoda festival based on division of tasks.

With respect to living conditions, one aspect has to be mentioned here, too: Some villages are very united and engaged in carrying out village developments such as road constructions, establishing and renovating health centres and schools on a self-help basis.

---

11   Code Division Multiple Access

4.3   EDUCATION SITUATION

4.3.1 Education pattern in general

The education system in general is structured into primary school: from grade 1 to 5, middle school: from grade 6 to 9, and high school: from grade 10 to 11. Since 2012 all primary schools except the one in Nanpade have been upgraded to 'middle school branch', and the middle school in Main Naung has been upgraded to 'high school branch'. Such branches are not a school of that level with all grades. Moreover, they have only an upgrading of grades to some extent, which depends on its infrastructure and manpower. In the way of teaching and of administration the branch depends on its relevant official school. Therefore, high school level education was available in Loneton and Main Naung, even Shweletpan and Hepa could offer only the first half of middle school level, all the rest villages offered education at middle level except Nanpade, where only primary level was available. Mohnyin (the nearest town outside the research area) has one Degree College and one Technological University. These institutions offer the university education level for the research area. For instance, currently (2015) 138 students (108 distance education students, 30 day students) from the Indawgyi Lake Area are enrolled at the Mohnyin Degree College in the first year (IDGY-45). But some subjects such as Law, English, Geology, Archaeology and Economics are only available in Myitkyina University. According to information of respondents, the average distance to primary school, middle school, high school and university in the area is 0.7 miles, 1.4 miles, 6.4 miles and 35 miles respectively.

Table 4.6: Number of teachers and pupils in the research area

Village	Primary school			Middle school		
	no. teacher	no. pupil	teacher/ pupil	no. teacher	no. pupil	teacher/ pupil
Nyaungbin	7	303	43.3	8	274	34.3
Nyaungbin (1)	10	284	28.4	1	28	28.0
Nammilaung	8	130	16.3	5	122	24.4
Nanpade	7	135	19.3			
Loneton	4	168	42.0	9	182	20.2
Loneton (Susi)	2	43	21.5			
Mamomkai	9	309	34.3	5	201	40.2
Main Naung	21	328	15.6	8	533	66.6
Main Naung (Botegone)	7	189	27.0			
Main Naung (Susi)	9	221	24.6			
Shweletpan	3	87	29.0	2	44	22.0
Hepa	5	141	28.2	1	34	34.0
Nammoukkam	10	338	33.8	3	203	67.7
Lonsent	4	184	46.0	7	168	24.0

As can be seen in table 4.6 the teacher to pupil ratio partly varies remarkably between the schools and villages. The primary school teacher to pupil ratio was quite problematic in Lonsent with 1:46, in Nyaungbin with 1:43, and in Loneton with 1:42. In the other villages it varied from 1:34 to 1:16 respectively. On the middle school level a high variation could be found, too. The ratio was problematic with 1:68 in Nammoukkam, in Main Naung with 1:67, in Mamomkai with 1:40. In the others villages the ratio varied from 1:34 to 1:20 respectively. High schools (not shown in the table) are only in Loneton and Main Naung. The ratios vary less and are quite reasonable with 1:25 in Loneton and 1:15 in Main Naung.

Schooling status

60% of total household members already left school, 32% were attending school and 8% never attended school.

Currently attending school: 38% of the household members, who currently attend school, are primary students. 31% are middle school students and high school students rate 13%. University students contribute with 10%. In this case two types of students were found: one was day-university student[12] with 4.3% and the other is distance-university student[13] with 5.8%. Nursery constitutes 8% attendance.

Never attending: Table 4.7 shows that in the older age groups a higher percentage of people have never attended any school. This result is not surprising, because 'their school time' (around 1940s) is in the pre-independent period of Myanmar, when the political situation was unstable and a lack of transportation was widespread. So, children had difficulties to have access to school. Obviously, the second largest group (in percentage) was formed by the infant group (0–5 years old). Common in Myanmar is that children when five years old are allowed to go to school. Interesting is that 5% of the age group between 21–25 years old never attended school, but the significant reason for that was not evident, it could be coincidental. Figure 4.2 also shows that the first reason why people older than five years (n=23) never attended school was 'school too far' with 39%, and the second reason represented 'need to work' with 22%. The other reasons such as 'costs', 'sick', 'disable' and 'orphaned' almost equally rated with about 9%.

Regarding the illiteracy issue, 1.8% of the total household members are classified as illiterates. Beside also 0.4% of total household members are not able to read. One reason is, for two and half decades the literacy programme has not existed anymore in the region.

---

12   Student has to attend the class for the whole semester.
13   Student just attends max. 2.5 months or min. 2 months: depending on the subject for one semester, parallel can job.

Table 4.7: Relation between age group and never attended & left school

Age group	Never attended school		Left school	
	Absolute frequency	% frequency of age group	Absolute frequency	% frequency of age group
0–5	77	63.6	3	2.5
6–10	1	0.7	2	1.4
11–15	0	0.0	13	8.7
16–20	2	1.4	78	55.4
21–25	6	5.0	101	84.2
26–30	0	0.0	108	100.0
31–35	1	1.0	95	99.0
56–40	0	0.0	74	100.0
41–45	1	1.4	71	98.6
46–50	0	0.0	53	100.0
51–55	1	1.4	69	98.6
56–60	6	13.6	38	86.4
61–65	0	0.0	18	100.0
66–70	4	28.6	10	71.4
71–75	3	30.0	7	70.0
76–80	0	0.0	7	100.0
81–85	2	66.7	1	33.3

Figure 4.2: Reason for never attending school

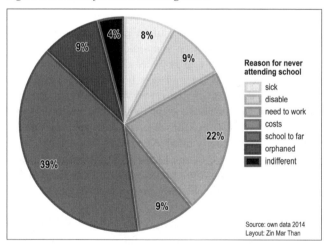

Left school: Table 4.7 shows that household members already have left school in their age group 16–20 with 55.4%. This high percentage is not a surprise when keeping in mind, that every village does not have access to high school. If such schools are not available in the village there are additional costs and transport is difficult in rainy season as well. These problems are precisely stated by an expert:

"If a student attends high school education in other place: Loneton, it costs 10 lakhs per school year, which is included for accommodation, food and study guide. So, most of the parents cannot afford it" (IDGY-12).

Another reason could be insufficient manpower and lack of teaching media at the school, which impact on matriculation exam, as some experts mentioned. One respondent also clearly pointed out:

"Loneton high school offers the high school education to all villages around the lake, I acknowledge very much how well the teachers take care of the students, but I cannot recommend the quality of teaching" (IDGY-09).

84.2% have skipped the school in their age group 21–25 and 100% quitted the school in their age group 26–30 (cf. table 4.7).

Figure 4.3 shows that the main reason for having left school was 'need to work' with 35% of total respondents (n=745), secondly, transportation was given as a reason with 21% and 'costs' played a third role with 18%. The other reasons such as 'completed desired level', 'indifferent' and 'got married' rated 11%, 8% and 6% respectively. According to the expert interviews, most of the experts (IDGY-12, 33 and 36) mentioned the same reasons, which showed by household interviews, namely that before 2004–2005 children were finished with the primary school and quitted the school due to either the difficult transportation or need to help parents or costs. These were also more or less the same reasons for never attending school. Some experts (IDGY-34 and 36) said that nowadays even though the situation is getting better in the area, some bottlenecks are still tangible, for instance:

"Only the primary education is available in the village, children can get middle school education within a distance of 2.5 miles, in rainy season it is very difficult to reach it" (IDGY-24).

Figure 4.3: Reason for left school

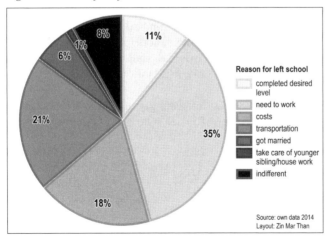

After middle school some of the students quitted school because they were not able to attend the high school education in other places due to costs (IDGY-34).

On the other hand, experts (IDGY-33 and 34) also highlighted the ongoing improvement. For example when Main Naung did not have a high school branch (no grade 10 available), 30% of high school students quitted school. But since 2012 the middle school was extended and offered grade 10 (high school education). So only 10% of the high school students quitted school due to the economic situation and the need to work. Moreover, parents, in particular young parents, are more interested for their children's education and strongly support it. Even if some parents cannot afford the cost for a high school student, they tried to send their children to the private school[14] in Nanma, Hopin, and Mohnyin with the credit/ instalment system because they understand very well that education is the only solution to lift the living standard. A village head explains about his interesting attitude:

> "I try to support my daughter, because I did not have this chance. I just finished with primary school...." (IDGY-24).

School level attained: The majority of the household members with 60% attained only the monastic or primary school level education. This high proportion of low-level school education is grounded in the situation of the past. One expert explains very clearly:

> "When I was in student age there was only available to go to the monastery, thus in our village people in the age of 40 and above just went to the monastery. At the time, the area was insurgent and the teacher could not come to the area. So, we could not learn properly" (IDGY-21).

A little more than a quarter of the household members (26%) finish middle school level education. It is also understandable because experts from Nammilaung and Nammoukkam (IDGY-16 and 23) said that today all students could finish their middle school education either in their village or in other village with a maximum distance of 2.5 miles.

Only very few people, 5.5% of the household members attain, a high school level education. It is not a surprise when keeping in mind all high school education bottlenecks esp. insufficient manpower in public school. As a result, it is very difficult to pass the matriculation exam and costs: i.e. students depend on private schools because high school education is only available in Loneton and Main Naung has high school branch.

University graduates constitute 8.5%. Interestingly the number of female university graduates is double the number of males who got a university degree (more detail in 4.1).

---

14 In the area private schools offer accommodation, food, and teaching. The parents have to pay for these services. The exams have to be done by public school.

4.3.2 Education patterns and socio-economic aspects

Relation between school level attained and age

The contingency table analysis proves that there is a highly significant correlation between school level attained and age. At C = 0.529, the contingency coefficient is quite high. Table 4.8 shows that an extremely high number of members who fall into the 'older' age group 35–44, 45–54, 55–64 and 65 and above having only a monastic or primary school level degree. By contrast, a very disproportionately low number of members from these have a middle school degree. Members of the age group 15–24 rated an extraordinarily high number with middle and high school level attained while there was also a remarkable low number with monastic or primary school level. Members of the age group 25–34 do have a very disproportionately high number of university degrees. The extremely low number of high school education level in this age group leads to the conclusion that in case of high school degree such students go a step further and aspire to university education.

Table 4.8: The relation between highest school level attained and age

School level attained	Age group					
	15–24	25–34	35–44	45–54	55–64	≥ 65
Monastic or primary	----	--	++++	++++	++++	++++
Middle school	++++		--	----	----	----
High school	++++	----	----	--		
University		++++		--	----	

Legend see table 4.2

Relation between school level attained and village

There is quite a significant difference between the villages regarding the school level attained. Table 4.9 shows that in Nyaungbin and Loneton a very disproportionately high number with university education level was found, while in Hepa such a situation can be reported for the number of members with monastic or primary level attainment. Respondents from Mamomkai and Main Naung consisted of a significantly high number with high school level attainment.

Table 4.9: The correlation between school level attained and village

Village	School level attained			
	Monastic/ primary	Middle school	High school	University
Nyaungbin				++++
Nammilaung				
Nanpade				--
Loneton	----	++		++++
Mamomkai			++	
Main Naung			++	
Shweletpan	++		--	--
Hepa	++++	--		
Nammoukkan				
Lonsent	++	--		

Legend see table 4.2

Relation between school level attained and occupation

The contingency table (cf. table 4.10) proves an extremely high number of government or private staff with higher education level, while a very disproportionately low number of farmers, fishermen and casual workers with higher education level is found. Persons from the fishing and farming sectors had extraordinarily often only a monastic or primary education, by contrast, an extremely low number of respondents from government or private and other (such as retire and broker) sectors attained only monastic or primary education level. It was also rare to find government or private staff with middle school level. Respondents from the self-employment sector such as retail shop, trade and restaurant showed no significant features with neither middle nor higher education level, but a disproportionately low number of respondents with monastic or primary education level is found.

Table 4.10: The correlation between school level attained and occupation

Occupation	School level attained		
	Monastic/primary	Middle	Higher*
Farming	++++		----
Fishing	++++		----
Self-employment	--		
Government/private staff	----	----	++++
Casual work			----
Other	----	++++	

* Higher = high school and university
Legend see table 4.2

The above results can be underlined by the results of table 4.11. It shows the percentages of school level degrees people in different occupations have attained. More than two third of the farmers have only monastic or primary school education. And in fishing this figure is even higher (more than three-quarter). The only occupation sector in which many people have a higher education is the government or private sector in which more than four-fifth has at least a high school degree. In the sector 'other self-employment' (mainly shopkeepers and mine owners) a little less than 50% have only monastic or primary school education. But in this sector there is quite a high portion of middle school education (more than a third).

Table 4.11: Percentages of attained school level in different occupations

	School level attained		
Occupation	**Monastic/primary**	**Middle**	**Higher***
Farming	71.1	26.4	2.6
Fishing	76.4	21.4	2.3
Self-employment	48.6	36.2	15.2
Government/private staff	6.4	12.7	81.0
Casual labour	68.1	29.7	2.2
Total	60.4	26.4	13.2

* Higher = high school and university

4.3.3. Weaknesses and strength of the current higher education in the lake region

"[…] All villages need to upgrade the school level, because there have only primary or middle school, some cannot continue their study, even nowaday everyone should finish with high school education level. Thus, I cannot be satisfied with the current situation" (IDGY-09).

The lack of opportunity for high school education as just mentioned is also apparent at university level. Mohnyin Degree College offered only eight subjects such as Myanmar, History, Geography, Physics, Chemistry, Mathematic, Zoology and Botany. Students, who study other subjects, had to go to Myitkyina, Mandalay and Yangon, which cause costs. One officer of Mohnyin Degree College (IDGY-26) mentioned that students have a lack of chance to attend some other courses such as languages, computer programming, accounting, which students in urban areas can do. Moreover, students who study History, Geography, Zoology and Botany have a lack of regional base knowledge, which is still lacking at all universities in Myanmar. One expert assessed

"The quality of Mohnyin Degree College is not high/good enough, as we know everywhere in Myanmar, all are the same, […] It is far to become an academic person" (IDGY-08).

On the other hand, the experts pointed out very positively that in Mohnyin, which is not far from the research area, comparatively good opportunities are found in higher education. One expert assumed that

"It is very convenient for students to have access to higher education in Mohnyin, during weekend they can come back home by motor-bike" (IDGY-08).

4.3.4 Future development of the education sector in the lake region

Intended future initiatives

Regarding future initiatives for improving education, which are already in the state of concrete planning, three factors have been mentioned in particular:
(1) Upgrading the school and degree college level and increasing manpower: As an example an administration officer (IDGY-06) confirmed that the Ministry of Education intends to upgrade Mohnyin Degree College to Mohnyin University.

"[…] The infrastructures such as lecture rooms and student dormitory are ready. Coming budget year 2014–2015 we will be ready to get Mohnyin University. Moreover, we have already reported to relevant Ministry. Indawgyi region can benefit from it" (IDGY-06).

Such a university would offer all subjects, so that all students would not need to travel far and for parents the cost would be lower. If studying nearby the chance of staying for work in the area after study is assumed to be high. But, at least in 2015 when doing the second phase of the field research this upgrading has not taken place.
According to interviewees (IDGY-12, 16 and 36) there would be potential to upgrade the branches to official schools (i.e. middle school branch would become middle school and high school branch would become high school) especially in Shweletpan, Nammoukkam, Lonsent and Main Naung. A village head (IDGY-24) noticed that Nanpade also is trying to upgrade the primary school to middle school branch. Parallel the number of teacher could increase as an administration officer stated:

"In coming year (2014–2015) the number of teacher will increase" (IDGY-07).

(2) Improvement of infrastructure especially the road condition: A head of a school supposed that a good transportation could be part of the future opportunity for education.

"Now the road is being built, if the road can be used for the whole year, it would be good for the health and education as well" (IDGY-16).

(3) New curriculum of the school: It is confirmed by the majority of the respondents, that in the upcoming school year (2014–2015) students could learn ethnic language at school. Moreover, an international expert discussed environmental education courses based on Indawgyi Lake to complete the school curriculum. It could be one future opportunity of education.

Future needs for improving education in the lake area

As the majority of the respondents mentioned two main factors were essential for education development in the village, namely improving the general knowledge of local people, and enhancing the quality of education. In particular, education programmes are required for health, drugs and environmental awareness as well as vocational trainings for agriculture, education, handicraft and eco-tourism.

For example, one expert (IDGY-08) suggested that education programmes should focus on environmental awareness, and sustainable use of resources. Another respondent (IDGY-22) recommended health education programmes especially on the drug issue. In fact, 'how to organize to get people's participation for the programmes' was most important to consider. Therefore, an administration officer (IDGY-07) confirmed that changing the attitude of people would take time to understand or to participate. An international expert also stated that the most difficult thing in a developing process is always education, whereas implementing technology and infrastructure is definitely easier.

Regarding vocational training a school head said that to improve the quality of education a vocational training for teachers is essential, since some teachers are young and less experienced (IDGY-12). For instance, in some villages due to insufficient manpower volunteers, who finished with high school education level, helped out on the teaching. One respondent (IDGY-53) explained that vocational trainings for "how to do added value with regard to forest products and raw materials such as jade and gold" are necessary, because these are still lacking. An interviewee from an International None Government Organization (IDGY-11) suggested that an education centre for Indawgyi Wildlife Sanctuary should exist, which supports good insight for the wildlife but also for problems. To inform visitors and local people is very important. For instance, local people should be aware of the tourists and know what the limitations are, and what eco-tourism means.

Moreover, one interviewee (IDGY-27) supposed in order to improve the knowledge of people a library should be established in the village. To enhance the quality of education the level of school have to be upgraded and the school curricula need to be revised, e.g. middle school to high school and adding environmental education based on Indawgyi Lake in the curriculum. An officer from Mohnyin Degree College (IDGY-26) commented to upgrade the Degree College to University, then to set up a research unit based on regional issues, which does not exist in Myanmar yet, to develop the motivation of the students by learning or working together with national and international scholars or researchers and to keep the educated people for the region. In an interview with an international expert (IDGY-26) there was agreement with these comments and he remarked that university can help giving people option and knowledge. So, the higher the education standard, the better the whole situation is becoming. Moreover, it has one more side effect, too. If people have good job opportunities, they do not engage in conflicts, because everybody wants to have a good and secured life. The more stable the whole economy is, the more options are there and the less people want to have conflicts. Thus, good education is also part of conflict solution. A conservation officer (IDGY-05) also mentioned that staff education programme workshops, more meetings, more discussions are appropriate to solve the conflict between the local people and conservation.

4.4   HEALTH SITUATION

4.4.1 Current situation

The research area (the 10 surveyed villages) had a total population of about 19,000. But it had only one 16-beds hospital with one medical doctor and two nurses located in Loneton. The other villages except Shweletpan, Nammilaung and Nanpade had health centres with a midwife and in Nammoukkam with a nurse. The villages, which have no health centres, had access to a health centre within a maximum distance of three miles. The nearest other hospital is in Mohnyin with 100 beds and a total manpower of 133 including specialists, physicians, different levels of medical staff (e.g. doctor, nurse, midwife and so on), all levels of administration (from head of the hospital to guard) and staff from service departments in particular for sanitation (ironing, cleaning, etc.) (IDGY-09). The maximum distance from the villages of the research area to Mohnyin is 49 miles and the minimum distance is 29 miles, not considering the not optimal road condition, though it has been upgraded to some extent in 2015. An administration officer from Mohnyin noticed:

> "[…] Actually, Mohnyin hospital is not enough for the whole region. I heard that sometime some patients are put on the corridor" (IDGY-06).

According to the household interviews 77 members of 64 households, which makes 6.2% of the total members were sick during the last 60 days. The majority caught a common cold and it took from one to two weeks to recover. Very few respondents suffered from chronic illnesses such as hypertension. 40.3% of total patients went to a health centre, 36. 4% bought medicine at a drug shop themselves, 16.9% saw physician at a hospital, the rest of 6.5% visited a shaman. Health centre and hospital are public, therefore patients are charged for medicine and not for doctoral treatment (in private clinic and hospital both are charged).

The experts (IDGY-04, 22 and 35) often mentioned that local people have a lack of health knowledge, and many of them still believe in traditional healing practices, the shaman, and superstitiousness. The following situations are reported as examples: when people have fever, they smear oil on the neck and back, then they scratch these areas with a spoon, the body gets heat and the blood circulation gets better. They feel better immediately. But as a side effect they can get skin infections. When this happens, they go to the health centre for a proper treatment. Moreover, in the village some untrained or unofficial medicine men are working. They treat the patients quite often in an unsustainable way (i.e. they just focus on quick recovery, not thinking of side effects). Most of the patients like this way, because if they recover quickly they can work again. An expert from Nyaungbin noticed:

"The villagers still believe superstitiousness, for instance, there is recently a case: a child
suffered from canine madness and died, the villagers believed that it was superstitiousness and
it is caused by spirits. Even though the health education programme was held one month ago,
their believes still root in their mind" (IDGY-22).

Malaria was prevalent in the rainy season in the whole area although an enormous
high percentage (97.2% of total household members) used mosquito nets. The
reason for this discrepancy could be either improper use of the mosquito net or the
answers were inaccurate. An expert from Nyaungbin mentioned,

"There are about 15 to 20 patients (esp. farmer, mining worker, timber/firewood collector) per
month in the rainy season (July, August September) in the village, it is serious and the virus
spread in brain, but not deadly. In dry season there are not many and only one or two patients
per month in Nyaungbin" (IDGY-22).

She also mentioned that some villagers do not know that malaria is a vector
(mosquito) borne disease and still have their traditional beliefs, namely that drinking stream water or eating banana, causes it.

The second common disease in the area was diarrhoea in the rainy season and
flu in the dry season. Experts from Hepa, Shweletpan and Nyaungbin indicated that
the hygienic status is very low in the villages; for example: six households in
Nyaungbin do not built a latrine and share with the neighbours. Beside these
diseases the following health matters have been named: malnutrition, abortions,
hypertension, hepatitis B and C, kidney problems, road traffic accidents (lower in
rainy season but deadly and high in dry and summer season) and work place acci-
dent in gold mining (8 people died in 2012 (IDGY-35)).

Additionally a few HIV patients were reported in the research area, due to drug
abuse. An expert from Nyaungbin said about drug abuse:

"The chemical drug is prevalent in the area and the users either inject it into the blood or inhale
it. If so, I guess the user can survive only for two years. In rainy season 2013 from three to five
users (between 20 and 30 age) died per month in the village due to drug abuse" (IDGY-22).

The experts (IDGY-04 and 19) also complain about the absence of proper storage
rooms for medicine in the whole research area (esp. vaccines). Additionally a just
minor availability for an ambulance vehicle has been reported. An expert described
one of her experiences as:

"If there is serious case, patient is poor and has to go to hospital, then collect the money in the
village and look for a car, when the patient arrived at hospital, it is late"(IDGY-17).

According to the experts (IDGY-19, 22 and 35) since 2011 the government and
non-governmental organizations supplied some medicines, so that they were free
of charge and enough for child, senior and maternal people. In addition social
associations carried the travel costs and food for those who were poor and stayed
in the hospital and a very few villages owned community cars which could be used
as an ambulance. Since 2011 experts (IDGY-22, 26 and 35) organized the health
education courses on topics like malaria, diarrhoea, and drug issue at least once per
year in the villages, at the school, at the Mohnyin Degree College. Besides that, all

midwifes tried to give the health education with regard to nutrition, family planning, and personal hygiene to the villagers whenever they do have a chance.

4.4.2 Future development of the health sector in the lake region

Intended future initiatives

An administration official from Mohnyin (IDGY-06) informed that in the near future improvement in the health infrastructure would take place. For instance, the Mohnyin 100-beds hospital will be upgraded to 200 beds, and one traditional medicine hospital will be built in Mohnyin by the government. Moreover, according to 2014–2015 budget allocation the health sub-centre in the village of Hepu will get additional personnel (nurse or midwife) (IDGY-07). Another expert (IDGY-16) stated that the road is being repaired, so if the road can be used for the whole year, it can help to get a better health situation in the area in future.

The community participation is an important factor in contributing to the future opportunities of the health situation in the area. For example, an expert from Nanpade, where a health centre does not exist until yet, explained, that the village community planned to construct a health centre by its own and will apply for personnel (midwife or nurse) from government. In addition, communities quite often gave assistance to the health staff, who need accommodation and they participated in health education and vaccination programme. Besides that, there are a few students in the villages, who currently study at the Nursing College; if they come back to the village and work there this will improve the health situation. Thus, all these mentioned factors will be tangible development potentials of the health situation in the area.

Most of the experts are concerned about HIV patients in the villages. Some patients are treated by social associations in Myitkyina, some are not. It is questionable how many patients exist, how to treat them and how to prevent this disease. Indeed, this will be an inevitable future threat due to lack of proper health facilities. Furthermore, some experts mentioned that the locals believe in shamans, and the unsustainable treatment way used by untrained or unofficial medicine men are also difficult to overcome.

Future needs for improving the health sector in the lake area

The majority of the experts confirmed that for improving the situation health education seminars (esp. on the drug issue) and local people participation are first priority. Second priority are health facilities i.e. every village should have a health centre with midwife or nurse and easy access to ambulance service, and a sufficient opportunity for medicine transport. A third issue mentioned is to control the medical system by the local authority, in particular in such a way that the untrained or unofficial medicine men in the village are not allowed. Moreover, the experts from

Mamomkai and Nammilaung recommended that people need to change their life-style and mindset, and then to visit the health centre is also necessary. For instance, a school head mentioned that a sport programme should be installed at the school.

4.5   MIGRATION TRENDS

4.5.1 Socio-demographic structures of immigrants

According to the survey 10% (n = 21) of the interviewed households had immigrated into the research area during the last ten years. 38% of these cases moved from Middle Myanmar, 29% from Mohnyin Township, 19% from places within the Kachin state, 10% moved from directly neighbouring villages, the rest of the immigrated households were from other places in the country. 'Work' was by far the main reason for households to immigrate (95%). Other reasons are named rarely, e.g. marriage. Since 2010 the immigration rate in the area increased with the years, the peak year of in-migration was 2013 with 43% of all immigrated households, and 2012 was the second highest recorded year with 14%. In the earlier years from 2005 to 2009 the immigration rates were recorded with 5% per year.

The immigrated households consisted of altogether 68 members, which represented 5.5% of total population of surveyed households. 32 of these members are female and 36 are males. Table 4.12 shows to which occupational group these members belong. The majority of the household members was employed in fishing with 23.5% and the government or private staff sector, which contributed 20.6%. Very few household members worked in farming with 2.9%. The numbers show obviously that in-migrated families intended to work for fishing and in the government or private sector in the area. The non-occupational groups such as students and dependents also contributed highly with 17.6% and 23.5%.

Table 4.12: Occupation and non-occupation of immigrated household member

	Absolute frequency	Percentage frequency
Occupation		
Fishing	16	23.5
Government/private staff	14	20.6
Self-employment	4	5.9
Farming	2	2.9
Retail shopkeeper	2	2.9
Casual work	1	1.5
Other	1	1.5
Non-occupation		
Student	12	17.6
Dependent	16	23.5
Total	68	100.0

Regarding the immigrated household income a minimum of 1.2 Mill. Kyat (nearly
€ 923) and a maximum of 8.8 Mill. Kyat (approx. € 6,769) had been reported,
resulting in an average income per household of 3.4 Mill. Kyat (nearly € 2,615) and
a positive balance per year with an average of 1.7 Mill. Kyat (nearly € 1,307) except
one household, which consisted of some students. Although the averages for
income and balance are lower than the averages of all interviewed households (see
in more detail in subchapter 6.2) – income (4.9 Mill. Kyat (approx. € 3769)) and
balance (2 Mill. Kyat (approx. € 1538)) – it can be stated that almost all immigrated
households had quite a reasonable income/balance because the minimum income is
distinctly higher for them than for all the households, which is only 0.2 Mill. Kyat
(approx. € 154).

According to the table 4.13 the majority of the immigrated household members
attained monastic or primary school education level with 49% (n=53), 30% had a
university degree. Compared to the numbers related to the total household members
the proportion of people with a university degree is quite high whereas the proportion with a low education level is considerably low. This for a first glance positive
situation gets a more realistic perspective when the relation between occupation and
education is considered.

Table 4.13: School level attained by immigrated and total household member

School level attained	Absolute frequency	Percentage frequency	Absolute frequency of total HH member	Percentage frequency of total HH member
Monastic/primary	26	49.1	565	59.9
Middle school	10	18.9	246	26.1
High school	1	1.9	53	5.6
University	16	30.2	80	8.5
Total	53	100.0	944	100.0

Relation between occupation and education level attained by immigrated
members during last ten years

The contingency table analysis (cf. table 4.14) proves a significant correlation
between occupation and education level of immigrated household members. An
extremely high number of fishermen with low to medium education level and by
contrast a remarkable high number of government or private staff with higher
education level was found. Moreover, the 'Rest' (represents farming, retail shop-
keeper, self-employment, and casual worker) has no disproportional figure with
neither low to middle nor higher education.

Besides the kind of migration mentioned above the type of seasonal migration
was frequent and consisted mainly of fishermen and gold mining workers. Migrants
were mostly between 16 and 40 years old.

Table 4.14: Relation between occupation and education level by immigrants

	Education level attained	
Occupation	Low to middle	Higher*
Fishing	++++	----
Government /private staff	----	++++
Rest		

* low to middle = monastic or primary and middle school
* higher = high school and university
Legend see table 4.2

According to information from the experts (IDGY-21, 31 and 34) the highest number of migrating fishermen was recorded in Nyaungbin with 50 households every year. In other villages namely Loneton, Hepa, Nammoukkam, and Shweletpan about 15 to 20 migrated fishermen households were counted. This migration trend has developed over the past 15 years. Mostly these seasonal migrants were from Sagaing Division and Mandalay Division (Middle Myanmar) and a few from Shan state.

Experts from Loneton, Nyaungbin and Shweletpan (IDGY-21, 31 and 36) explained that seasonal migrants are living with a guest registration[15] in the village, one of the migrants is selected as a leader, who has to communicate with the village head about institutional issues and also takes the responsibility for all migrants. An administration staff from Shweletpan (IDGY-36) stated that seasonal migrants have to pay 5000 Kyat per household per month to the village for guest registration, and they have to pay 20.000 Kyat per year to the village if they want to build a hut at the bank of the lake.

The seasonally migrating fishermen lived on the banks of the lake, which belong to the lake conservation zone. One interviewee from Shweletpan also complained about the problem that migrants are using the very small mesh size fishing gear and fish throughout the year. Moreover, another expert explained:

> "[…] We are not satisfied with their litter and hygiene at the bank of the lake, even they are in harmony with the local […] We want to offer them to buy the land and settle in the village, as they are seasonal migrants since 15 years ago" (IDGY-36).

An administration official from Loneton (IDGY-31) explained that in Loneton migrants can buy the house but they do not possess the land.

Near Main Naung village and Nyaungbin, where some gold mines are located, the seasonal migrating gold mining workers were living with 20 households in Main Naung, and 10 households in Nyaungbin, and some others lived in the mining areas (IDGY-34). The seasonal gold mining labour migration trend started in 2000. The majority of the migrants were from Sagain Division and Mandalay Division and a few from Rakhine State. One interviewee from Nyaungbin complained,

---

15   Non-local has to inform village administration office for the stay.

"[...] The migrants are not in harmony with the local people, they are rude and discriminate the
local people" (IDGY-21).

The situation in Main Naung was different from Nyaungbin. One respondent stated,

"[...] There is no conflict between local and migrants" (IDGY-32).

Regarding the seasonal migrants in Nyaungbin an administration official explained
that some migrants wanted to work at jade mining in Pharkant, but they did not
have a chance to stay there then moved to Nyaungbin. At first they arrived indivi-
dually and one or two years later they brought their families and now they stay in
Nyaungbin permanently. The situation of Main Naung seemed to be the same, an
administration official from Maing Naung explained,

"[...] Migrants, who are seasonal and stay with temporary family registration during first one
and two years, then they buy the land and settle in the village, here is easy to have access to
timber, bamboo and land" (IDGY-34).

One retired teacher mentioned that this village, Main Naung, was established in
1946. It was homestead for about 60 households in 1961. In 1963 25 Kachin households, who lived on the mountains and depended on shifting cultivation, were
relocated to Main Naung, today the village has 1087 households. So, almost all
villagers are migrants, but there are no seasonal migrating fishermen, because the
lake is a bit far from the village. An administration official from Main Naung also
confirmed that every year at least 20 households immigrated due to job opportunity.
According to the information from the experts (IDGY-21 and 34) the highest
number of immigrated households were found in Main Naung and in Nyaungbin,
which is also closer to Hpakant jade mining. The village head of Nyaungbin (IDGY-
21) informed that the immigrated household was 10% of total households (n=529)
in the village including seasonal fishermen and gold mining workers household.

4.5.2 Socio-demographic structures of emigrants

Sixty-six household members, who have emigrated during last ten years, were
recorded in the household questionnaires. The emigrants consisted of 34 males and
32 females and represented 5.3% of the total surveyed population. The majority
moved to other places within the Kachin State with 30%, within the Mohnyin township and to middle Myanmar the emigrant figures rated to 25% each. As the main
reason of emigration 'work' was reported with 50% and 47% of total respondents
confirmed the reason 'education'. The recorded number of emigrants took place
from 2006 to 2013. The peak year of emigration was 2013 when 49% of the
emigration happened. Additionally, it has to be stated that the emigration rate
constantly increased over the years.

Table 4.15 shows the education level of the emigrated household members, the
majority attained middle school and high school level with 36.5% and 34.9%
(n=63), and for 19% of the emigrants a university degree is recorded. Emigrated

members with monastic or primary education level contributed really low with only 9.5%.

Table 4.15: School level attained by emigrated and total household member

School level attained	Absolute frequency	Percentage frequency	Frequency of total HH member	% frequency of total HH member
Monastic/primary	6	9.5	565	59.9
Middle school	23	36.5	246	26.1
High school	22	34.9	53	5.6
University	12	19.0	80	8.5
Total	63	100.0	944	100.0

Table 4.16: Occupation and non-occupation of emigrated household member

	Absolute frequency	Percentage frequency
Occupation		
Government/private staff	14	21.5
Casual worker	11	16.9
Self-employment	4	6.2
Retire	2	3.1
Trade	1	1.5
Other	1	1.5
Non-occupation		
Student	32	49.2
Total	65	100.0

The affiliation of the emigrated household members to the occupational and non-occupational groups are shown in Table 4.16 The highest number of emigrated persons was employed in government or private sector with 21.5%, and casual workers built the group with the second highest percentage (= 16.9%). Only students belong to the group with non-occupational status, and they were the majority of emigrated members with 49.2%.

One of the school heads (IDGY-16) highlighted that the students did not return to the area after they have been educated but got a job somewhere in Myanmar. Insofar the area suffers a brain drain. Moreover, one expert (IDGY-12) also commented that the area contains a lack of job opportunities except as teacher for the graduated people. Thus, the well-educated young people have to leave the area and work in other parts of Myanmar and abroad. According to the discussion with the experts, the solution for preventing the brain drain is to create adequate job opportunities in the area. Additionally, an interviewee, who is a local, and a master student at the University of Education, complained:

"After my study, I will work as a teacher, so, I need to know more knowledge than the students, here lack of information in the village, therefore my knowledge will be limited here and will not be developed like the teachers are in other part of Myanmar" (IDGY-16).

Relation between occupation and education level attained by emigrated members during last ten years

According to the contingency table (cf. table 4.17), there is a pronouncedly clear relation between the occupation level and the education level attained for emigrated persons.

Table 4.17: Relation between occupation and education level by emigrants

Occupation	Education level attained	
	Low to middle*	Higher*
Government / private staff	----	++++
Casual worker	++++	----
Rest		

* Low to middle = monastic or primary and middle school
* Higher = high school and university
Legend see table 4.2

An extraordinarily high number of emigrants attained higher education are government or private staff, by contrast, a very high number of emigrants with low to middle education level are engaged in casual work. The occupational category 'Rest' (e.g. self-employment, retire and trade) does not show any disproportional figures with regard to education.

4.5.3 The balance of migration

In general, the migration process leads to a situation, which can be summarized as follows: beside staff from the government sector mainly unskilled and business people, who only focus on income, immigrated into the area, this led to some social and environmental problems. The emigration process consists of well-educated young people to a great extent and results in particular in brain drain, which constitutes a major challenge for the area.

5   INFRASTRUCTURE SITUATION

This chapter is dealing with different aspects of the infrastructure. Firstly, the situation of transportation (e.g. road conditions, transportation modes) is described. Subchapter 5.2 is focusing on communication and information elements. The electricity supply situation is the topic of subchapter 5.3, followed by a description of the water supply conditions. The chapter closes with subchapters on education facilities and on health facilities.

5.1   TRANSPORTATION

The lake area is connected to Hopin and Mohnyin by a motor road, which also leads around the lake almost entirely. In the past, this road was a two-way dirt road and was intensively used. During the rainy season from July to mid of September, when there is heavy rainfall, this dirt road regularly became muddy and sometimes impassable. To facilitate the transportation of vital goods, roads often needed to be repaired on a self-help basis by the local communities. The inter-village connection roads were mostly used by private people for interchange of goods and the sale of own agriculture products, fish, firewood and other goods, using cars, small lorries/tractors, bullock carts, motorbikes, bikes. A second equally important mode of transportation besides the sometime impassable roads was the waterway (by boat or motorboat).

There is also a once a day connection from Nyaungbin to Mohnyin and Hopin, operated by privately owned cars. From there access to rail and road transport to other parts of Myanmar is available. Every morning at a fixed schedule a car is collecting passengers to transport them to their destination for a fixed fee. However, in the past during the rainy season it was very difficult and took a long time, because the route crosses a mountain area and the road was a dirt road. The result was that people mostly depended on motorbike transport to reach Mohnyin and Hopin.

According to the household questionnaires, in every village the ownership rate of motorbikes was dominant over all other transportation items, like bicycle, motor vehicle or bullock cart. The highest ownership rate of bicycles was in Hepa. None of the inhabitants from Main Naung however owned a motorboat or other vessel, while none of the households from Nanpade, Mamomkai, Shweletpan and Hepa possessed a motor vehicle.

According to the questionnaires, 44.7% of the households evaluated the current transportation facilities as 'tending to bad', 37.2% rated them as 'tending to good', 15.3% said 'bad' and only 2.8% evaluated them as 'good'. Experts (IDGY-17 and 19) also mentioned that children quitted the school early as a consequence of transport bottlenecks. Until last year (2013) it was difficult to reach the hospital in time, thus the mortality rate was high. However, since late 2013 the situation was getting better, in the 2013–2014 budget year, the Hopin-Nyaungbin road (40 miles) to the west of the lake started to become upgraded to a paved road funded by Public Works, Ministry of Construction (see figure 1.2). In parallel at the east of the lake between Nantmon and Lonsent (14.5 miles) a 9.5 miles gravel road is also constructed, funded by the Ministry for Progress of Border Areas and National Races Development Affairs and implemented by the City Development Committee.

An administrative officer (IDGY-06) mentioned that in 2015 the Rural Development Department will also support the road construction, so, all together three parties will be involved in road improvement. Thus, in the 2014–2015 budget year the road around the lake will become a gravel road completely. The officer also informed that transportation is the 3rd priority for improvement of infrastructure in the area and that the five-year plan depends on the annual budget and the road will be upgraded gradually year by year. It is planned that in the coming five years the road around the lake will become a rural tar road. This generally positive outlook is underlined by the local people's assessment, which they gave in the household questionnaire: the future village transport facilities are evaluated positive by 68.9% and very positive by 15.3%; only 13.9% see it negatively and 1.9% has a very negative view. The Administrative Officer also mentioned:

> "[…] In the next five years, people in the Indawgyi Lake Area will have the same right regarding socio-economic status such as education, health care, infrastructure like urbanites do" (IDGY-07).

In 2015 the 40 miles Hopin-Nyaungbin paved road has been completed as well as the 14.5 miles Nantmon-Lonsent gravel road. As a very positive consequence it can be stated that the road now can be used all year round, which improves the economic, health and education situation in the area significantly.

5.2   COMMUNICATION AND INFORMATION

There is only one public telegraph office in the entire area, which is located in Loneton. In 2008 a CDMA network was set up for the area (Zin Mar Than 2011). At that time a SIM card for this network cost between 900,000 Kyat (approx. € 692) and 1,100,000 Kyat (approx. € 846). In 2013 the government provided very few CDMA and GSM SIM cards at 1500 Kyat (approx. € 1.15) for government staff in the area, but this was insufficient. Since October 2014 GSM SIM cards are provided by private companies and are available for everybody at 1500 Kyat (approx. € 1.15). The number of mobile phones, line phones and the percentage of households who own one or more were recorded in the villages in January 2014 (more detail in 4.2). In general, the ownership of mobile phones had increased significantly in the last couple of years. The highest dispersal rate of mobile phones was in Loneton, whereas the possession of line phone in the area is almost zero. An exception was in Nammilaung and Loneton, where four accesses could be recorded.

However, according to field observation poor connections can be noted which is also confirmed by the evaluations, which were done by questionnaires, in which the current communication facility counted 'tending to bad' with 48.6%, of surveyed households, 12% evaluated 'bad', 36.6% commented 'tending to good' and very few 2.8% said 'good'. Nevertheless, one expert (IDGY-10) stated that the communication situation in the region has developed since 2012, and better communication leads to more economic development. According to the local assessments the communication facilities in the future were rated 'positive' by 67.3% (n= 208), 14.4% projected very positively, 16.8% forecasted negatively and only 1.4% estimated very negatively.

As mentioned in subchapter 4.2 the main information source in the area were electronic media such as mobile phone, satellite television and radio, and a second important source of information was word of mouth. On third place was print media such as public and private newspapers and journals. However, in Nammilaung, Shweletpan and Hepa the ratio of information source ratio of electronic media and word of mouth was almost 50:50 and in Nammoukkam and Lonesant the electronic media source was only slightly higher than the source 'word of mouth', whereas in the rest of the villages it was obviously higher. As per household interviews, the local people evaluated the current information facilities as 'tending to bad' at 47.4% and 'tending to good' at 41.9%, at the same time 7.9% complained that it was 'bad', while 2.8% said 'good'. With regard to the future 69.4% positively assessed the future status of information facilities, 14.4% marked 'very positive', but 14.8% answered negatively while 1.4% rated it 'very negative'.

5.3   ELECTRICITY SUPPLY

As per January 2014 household interviews, access to electricity was mostly community and private based in the research region. There were a few exceptions as for instance, in some villages several private households used a generator for their own need plus they supplied households in their neighbourhoods. Only in Loneton a public generator had been installed by the government in 1996. It supplied electricity for governmental offices and street lighting from 6:00 to 9:00 pm. All the supply sources charged similar rates. The charges depended on the appliances the power is used for. For instance one electric lamp costs 3000 Kyat (approx. € 2.31) per month and power supply was only from 6:00 to 9:00 pm. Besides that, some households owned solar cells, in this way 75.9% of total households had access to electricity for lighting and the rest depended on candlelight. In Shweletpan, Nammilaung and Nanpade almost 50% of households depended on candlelight and the other 50% owned solar batteries, while the households in other villages depended more on private or community generator. Therefore, the local people evaluated the current situation as 'bad' at 51.2%, 'tending to bad' at 32.9%, 'tending to good' at 14.1% and 'good' at 1.9 %.

In terms of expert interviews, an administrative officer stated that the electricity supply has first priority in their area development plan. Then he explained:

> "The Ministry of Electricity is doing this using union budget, to provide access to electricity from hydropower to 11 village tracts in the Indawgyi region. Installation began in the west of the lake up to Nyaungbin and in the east of the lake up to Lonsent. With the 2014–2015 budget the electricity supply will be completed" (IDGY-07).

An expert (IDGY-07) recommended that, after electricity supply is completed, it should be considered to add value to local products in particular to agricultural products. For example, a good private or community rice mill can be set up, and better communication and information can be hoped for. As per questionnaires 62.7% positively assessed the future electricity system, 14.8% said 'very positive', by contrast 13.4% marked 'negative' and 9.1% answered 'very negative'. So, a little more than one-fifth of the interviewed households assessed their situation negatively. This high proportion is not surprising because every summer the electricity supply even in Mohnyin is insufficient due to hydropower shortage, as was mentioned by one expert (IDGY-06).

In 2015 Loneton, Mamomkai and Main Naung have already access to electricity supply and the supply system is under process for the rest of the research area.

5.4   WATER SUPPLY

The inhabitants had access to two types of water sources, which are surface water (lake) and ground water (tube well/pump). The data of water sources for drinking and household use and the distance to access to water sources were recorded during the household survey. Only very few households in Nyaungbin, Loneton and Shweletpan depended on the lake for drinking water, and all the rest used tube well/pump. None of the respondents used separate sources for drinking and household water except seven households in Mamomkai. Furthermore, all households in Nammilaung possessed tube well/pump at home, and 82% of households from other villages had it at home, while only 50% of households in Mamomkai had water sources at home (more detail in 4.2).

According to the questionnaires 61.6% evaluated the current water quality as 'tending to good' and 13.3% said 'good', by contrast 16.1% answered 'tending to bad' and 9.0% marked 'bad'. Some experts (IDGY-01 and 12) from Loneton and Lonsent also complained that the drinking water quality is not good enough. In terms of farm water availability some farmers complained that paddy farming only depends on precipitation. Altogether 47.5% evaluated the current farm water availability as 'tending to bad', 22.8% said 'bad', however, 25.3% answered 'tending to good' and 4.3% felt that the situation was 'good'. An administrative officer (IDGY-06) explained that the water supply in particular farm water and drinking water plays the second priority role in their area development plan. According to own observation, rainwater-harvesting practice is almost not in use in the whole area.

5.5   EDUCATION FACILITIES

In the area primary, middle and high school education were available, all schools and staff were public. In the past school infrastructures and teachers' accommodation for those who are not from the area, were often repaired and constructed on a self-help basis by the local communities and several donors, who were gold mine owners and felt a kind of corporate social responsibility for the area. In general, every village had more or less access to middle school level education. Some villages in particular Nammoukkam still not only need to extend the classrooms due to the increased number of students but also more manpower, teaching media and sport programme are necessary. The current education facility status, considering infrastructure and manpower, was rated by the interviewees as 'tending to bad' at 55.3% and 'bad' at 6.5%; conversely, 35.3% said 'tending to good' and 2.8% answered 'good'.

According to expert interviews, an administrative officer (IDGY-06) noticed that since 2012 the Kachin State government and members of Parliament started to support the school rehabilitation in the villages, where it is needed. Some experts (IDGY-16 and 36) also commented that the education facilities in the area are better developed than in the past. Regarding the assessment of the future status of education facilities, considering infrastructure and manpower, 59.8% answered 'positive' and 9.6% said 'very positive', however 30.1% estimated the future negatively and 0.5% forecasted it as 'very negative'.

An administrative officer (IDGY-06) informed that in the near future Mohnyin Degree College will be upgraded to university status, which supports the higher education for the area. The facilities such as lecture rooms and dormitories are already constructed. Moreover, in the future the education levels, i.e. the step-up from primary to middle school branch or from middle school to high school branch will be extended and more manpower will be provided according to the budget allocation of the State government.

5.6   HEALTH FACILITIES

In the area there was a 16-beds hospital/dispensary with one medical doctor and two nurses located in Loneton, but Shweletpan, Nammilaung and Nanpade had nothing, while the other villages had access to a health centre mostly with one midwife. Therefore, the whole area depended much on the 100-beds hospital in Mohnyin. Theoretically, all health infrastructure and staff are public. The infrastructure is often repaired and constructed on a self-help basis by the local communities and donors.

According to expert interviews (IDGY-04, 17 and 19) no proper storage room for medicine in particular for vaccine does exist in the area, and there is no ambulance. Furthermore, medicine transport from Mohnyin to the health centres in the villages was very difficult due to car shortage in the villages, although the government and some social organizations provided the medicines. The evaluation of current health facilities given by the households, considering infrastructure and manpower, shows that 44.4% said 'tending to bad' and 30.1% commented 'bad', while 22.2% answered 'tending to good' and only 3.2% assessed 'good'.

One midwife (IDGY-35) explained that since 2013 the government started to provide some funding for infrastructures in the villages, if required. For instance, recently X-ray equipment was provided for Loneton hospital. Moreover, social associations from some villages possessed cars, which can be used as ambulance. In fact, some experts (IDGY-07, 08 and 09) confirmed that the current health facilities are better than in the past. An administrative officer (IDGY-07) mentioned that in the near future the 100-beds hospital in Mohnyin will be upgraded to 200 beds (which according to government policies should be the minimum for a district hospital (IDGY-40)) and parallel to this, a traditional medicine hospital will be built in Mohnyin. It was planned to increase manpower within the 2014–2015 budget year, but this plan has not been carried out in time.

The head of the Myitkyina hospital also mentioned some plans for the Indawgyi Lake Area:

> "[…] In IDGY area there is the 16-bedss hospital in Loneton and Chaung Wa. Now the new 16-beds hospital is being constructed in Nanmon and one medical doctor was appointed for this hospital. In Lewmon one new hospital at the village tract level like in Loneton is being built. So, around the IDGY in the north Chaungwa village tract level hospital, a bit further south Lwemon village tract level hospital, in the middle Loneton village tract level hospital, near the mountain Nantmon village tract level hospital will cover for basic health care of the people, who live around the lake. But if people need to see a physician, they have to come to Mohnyin" (IDGY-40).

He also confirmed that in June 2015 one medical doctor and five nurses are appointed to the village tract level hospital in Loneton.

In the household interviews the future health facilities were fairly positively assessed by 46.4% and very positively by 3.8%. However, 45.5% of the households evaluated this aspect negatively and 4.3% very negatively which is not surprising because some villages urgently need to build a health centre and improve/increase the manpower.

6   ECONOMIC SITUATION

This chapter presents results and findings of the economic situation of the Indawgyi Lake Region and consists of three subchapters. Firstly (6.1) the current economic activities in the area are characterised in detail. Secondly (6.2) the financial situation of the households (seen as economic units) is described before thirdly (6.3) the contribution of single household members to the household income is of interest.

As already mentioned in subchapter 1.2: Compared to other states and regions in Myanmar like Chin State and Tanintharyi Region, the development of Kachin State is also lagging behind economically and this is true for the research area, too.

6.1   ECONOMIC ACTIVITIES IN THE AREA

Based on the classification used in the questionnaire the activities have been categorized into the following groups:
1. Self-employment
    a) Agriculture/Farming
    b) Fishing
    c) Retail shop
    d) Other self-employment
2. Dependent employment
    a) Government or private staff
    b) Casual work

In the research area, alltogether 403 main economic activities have been named by the 216 households. As 81 households relied on one main economic activity, 83 households worked in two and 52 households named three economic activities. From the perspective of the above mentioned activities the following results can be reported: (1) 39.8% of the households are engaged in agriculture/farming, (2) fishing is done by 25.5%, (3) 17.6% are engaged in retailing, (4) other self-employment is mentioned by 16.7%, (5) government or private staff is named by 17.1% and (6) casual work is an activity for 27.8% of the households. Beside their main tasks, households often are engaged in other minor activities and small-scale animal husbandry.

Agriculture

Agriculture is the major income generating activity in the research area. Paddy is the main cultivated crop, followed by peanuts and soy beans, which are mostly cultivated in the villages east of the lake. Most of the agricultural land is located in the lake's fringe areas and around the villages. West of the lake, the main agricultural practice is characteristically mono-culture (paddy). Because all farmers use the same variety, the amount of yield depends on the quality of the soil and whether there are differences in the weather (IDGY-24). There are no dams or canal systems to irrigate the fields (IDGY-21). This causes problems because in some years the rain comes too late after planting (IDGY-15). Additionally, the soil quality in the west of the lake is not very good for farming, so farmers have to use fertilizers (IDGY-24). The yield per acre with about 60 baskets is remarkable lower than those east of the lake, where farmers get about 100 baskets per acre (IDGY-34), largely due to the above mentioned quality of soil. Consequently, most of the farmers east of the lake namely in Shweletpan, Hepa, Nammoukkam and Lonsent, use a two-season crop rotating system by cultivating peanut and soy bean after paddy harvesting (IDGY-20) (cf. figure 6.1).

Figure 6.1: Agricultural seasonal calendar

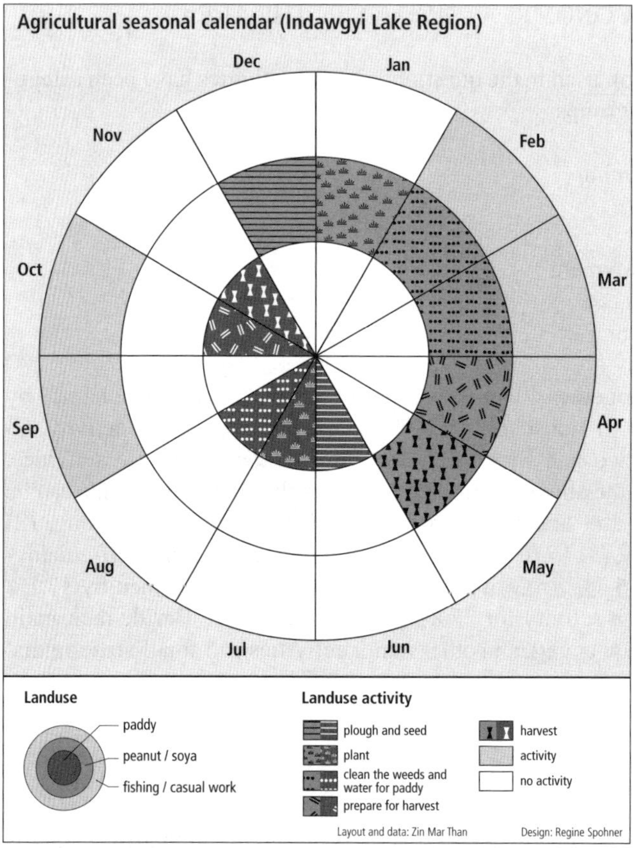

As figure 6.1 shows, the paddy farming starts in June and the ploughing and seeding last the whole month. Cultivating the fields with the plants from the nurseries starts in July, whereas in August the main work in the paddy fields is taking care of the water level and cleaning the weeds. In September no work has to be done in the fields, so that farmers undertake side jobs such as fishing or casual work, i.e. some farmers rent their bullock cart for firewood transportation, some operate in retail shops, work as brokers or as gold mining workers. In October the paddy harvesting is prepared parallel to side jobs, in November the paddy is harvested. After five to ten days of drying the paddy is threshed. If the soil is fertile enough, the land is ploughed for peanut/soy bean planting in December, and the actual planting takes place in January. In February and March the fields have to be cleaned from the weeds and farmers undertake side jobs again, in late April or May the peanut/soy bean is harvested. If no second crop is cultivated farmers stay in their other activities.

The above mentioned differences between the west and the east side of the lake due to soil quality can be added by another one, the difference in the size of the farms. There is one farm west of the lake, which is extraordinarily large with 265 acres. In order to keep it comparable this large farm is taken out from the further consideration of farm size. The boxplot of figure 6.2 and the parameters in table 6.1 show clearly, that the size of the farms in both the areas range over a more or less similar interval (west from two to 45 acres, east from three to 45 acres). But in the west much more farms are considerably smaller than in the east. In the west 50% of the farms (= median) have a size of less than nine and 75% have a size of less than 12 acres. In contrast, in the east the smaller half of the farms have sizes up to 14.5 acres, which is even higher than the third quartile of the farm size in the west. In the middle range (interquartile range = box) the farm sizes in the west stretch from 6.5 to 12 acres, so the middle 50% are concentrated in contrast to the middle 50% in the east, that ranges from eight to 22 acres.

Table 6.1: Parameters of the variable farm size: 'west' and 'east' of the lake

	Farm size in the west (acres)	Farm size in the east (acres)
Minimum	2.0	3.0
Maximum	45.0	45.0
Arithm. mean	10.3	16.6
1. Quartile	6.5	8.0
Median	9.0	14.5
3. Quartile	12.0	22.0
Stand. Deviation	7.2	10.9
Number of farms	51	38

Thus, the east-west differences in soil fertility, in agricultural activities and in farm sizes have strong effects on the household incomes earned in agriculture. On average, the total annual income from agriculture east of the lake is distinctly higher than in the west with about 5.1 Mill. Kyat (approx. € 3,923) compared to 2.1 Mill. Kyat (approx. € 1,615). The mono-culture system with one harvest per year on the western side leaches out the soil and this requires the use of frequently more ferti- lizers, which in return has adverse effects on the water quality and the eco-system (Chambers 1983: 86–87). Due to the limited crop production, most of the farmers work as fishermen after the harvesting, reinforcing the overfishing problem.

Figure 6.2: Boxplots: farm sizes: 'west' and 'east' of the lake

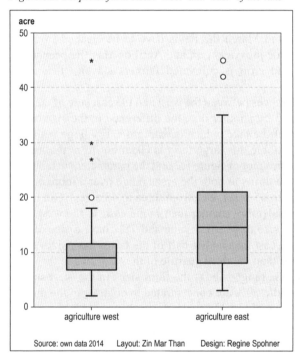

Most of the shifting cultivation lands can be found in the forest zone, which is conserved since 1999. Traditionally, mostly Kachin worked for it, and cultivated rice and pineapple (IDGY-32). The rice yield was about 25 to 30 baskets per acre (IDGY-34), but if farmers would take better care of the farmland the yield could increase up to 70 baskets (IDGY-30). Today, the government is organizing that the farmland in the forest/conservation area can be officially used by farmers in the near future (IDGY-34 and 31). However, until now there is no success, because of the unstable political situation. The Kachin Independence Army (KIA) occupies the forest and tries to impede the government's plan by preventing the local farmers from working in the forests (IDGY-44). One expert from the agriculture department of Kachin State states:

"Due to unstable political situation we cannot extend the farmlands, actually we have some areas to be used as farmlands, where there is no security, if we have peace we will extend the farmlands" (IDGY-39).

Nevertheless, according to two experts (IDGY-34 and 31) some wild land especially in Loneton (400 acres) and Main Naung (956 acres) is being allocated to villagers who do not own farmland but want to engage in farming.

"[…] In Loneton about 400 acres farmland, which is occupied by Military for years ago and yet it will be handed over average 10 acres per farmer, who does not own farm land and is interested to farm" (IDGY-31).

According to agricultural experts from Myitkyina (IDGY-39 and 52), the agricultural technique is still weak (especially quality species and knowledge of cultivation technique) and there are not enough agricultural specialists, who share their knowledge with the farmers. However, before and during the cultivation season agricultural specialists conduct knowledge sharing programmes in the township, where their office is located. But remote areas cannot be covered. How frequently workshops take place depends on the manpower. Even if some farmers understand and accept the new technique, they cannot follow due to the lack of resources (labour cost is expensive) and infrastructure. As a result, the yield is low. Additionally, most of the land is divided into small pieces and the land is not flat. Sometimes, their market attitude also contributes to the low yield, for instance:

"[…] Here the farmers want to cultivate the rice species which they prefer and their attitude is to fulfil their need and then they will sell the rest. Even our agriculture department advised to cultivate good yield species, but the farmers are not interested because they do not like to eat this rice, which is a bit hard" (IDGY-52).

Therefore, there are very few farmers who have agriculture knowledge and can invest in the area (IDGY-52). Since the year 2000 machines such as tractors, harvesting and threshing machines are used for farming in the area (IDGY-34 and 30). Today some farmers own machines on their own and some, who do not possess any, rent them from the owners, who usually collect charges for use. One expert highlights the pro and contra for this development:

"[…] Ten years ago the mechanized farming has taken place, as a result time consuming can be reduced, but disadvantages are less bio compost, use more fertiliser, the oil from machine can spill on the soil that leads to degradation of soil enrichment" (IDGY-18).

In addition, he also explains the current situation of livestock as some cattle is sold to China through the Shan State. There are two reasons to do so, namely the higher market prices stimulate to sell the livestock and the second reason is environmental:

"In the past, after farming season livestock could live freely and the owners collected them when they start to farm in May. But today, the owners take care of their livestock in order to law, which is organized by conservation team and local authority. Because the free living of livestock causes more consuming of bamboos and low yield [Farmers need the bamboo fence to protect their farm, otherwise livestock destroy the farm]" (IDGY-18).

However, some experts confirm (IDGY-28 and 30) that cattle and buffalos are still necessary for farming because some farmlands are muddy, wet, and upland, so only livestock can work on it. After doing the rough work with the tractor livestock can do the fine work and it can also be used for ploughing the corners of the farmlands as well as for making barriers to keep the water in the field (keep in mind that most of the farmlands are small). Additionally, livestock is very useful as a transport mode for firewood and agricultural products.

As mentioned earlier, in the area farmers own a minimum of two acres and a maximum of shortly over 45 acres. The medium is 15 acres for a farm. Since 2012 the agriculture bank lends money to farmers according to the following rules: only a farmer, who owns farmland, can get the loan of 100,000 Kyat (approx. € 77) per acre for max 10 acres (IDGY-28). Moreover, a farmer can buy farm machines with the instalment from the shop owned by government (IDGY-52). The interest was 1.5% in 2012 and has lowered to 0.75% in 2013 (IDGY-28). After harvesting the loan has to be repaid. One expert points out:

> "If a farmer owns more than 10 acres farmland, he can get only one Mill. Kyat loan. The rule is not proper for everyone who owns more than 10 acres. Today, the farming cost is getting higher and in reality, some farmers do not have enough money for farming. As a result, farmers have to presell their paddy to the brokers or traders, so they lost either 50% or 35% of their income [the preselling price is 50% or 35% lower than the market price after harvesting]" (IDGY-34).

The agricultural specialist from Myitkyina (IDGY-52) also confirms that the 100,000 Kyat (approx. € 77) per acre loan from the government during the farming season is not enough for the farmers. He suggests for improving the agricultural sector effectively a long–term loan, which should also be high enough, is necessary for farmers as well as an attractive and stable market.

According to experts (IDGY-30 and 23) since 2000 some farmers started to concentrate on gold mining and they reduce farming and just cultivate very little for their own needs. The lack of market accessibility made farmers worry that they neither can sell their products nor foresee the price. So farmers looked for alternative income opportunities in the area, which resulted often in gold mining. Reasons to change the economic activity were not only the lack of market accessibility but also some other factors such as insufficient farming income (low yield) due to lack of farming knowledge and infrastructures (esp. difficulty of farm water availability) and dependency on weather conditions (precipitation).

When gold mining started in the early 2000s (see subsection 'other self-employment' below) it was profitable. But over time people realised that the income generated by gold mining is unsustainable and gold mining has adverse effects on the (surrounding) environment. However, some people have already invested in this economic activity and as they are in debt, they cannot stop mining activities immediately. Nevertheless, since the introduction of a loan policy for farmers in 2012 as mentioned earlier and better access to market, people started to realize that farming is the best economic activity in the area. One agricultural officer points out:

> "[…] The yield totally depends on the weather, but the climate here is temperate and the area has enough rainfall, so here is better for agriculture compared to middle Myanmar, where it gets less rainfall, and delta region, where is flooded" (IDGY-39).

Today, two experts from Mamomkai village (IDGY-30 and 28) claim that gold mining affects farming in a negative way. For example, some streams near Main Naung and Mamomkai are deposited by mud produced in gold mines. When farmers irrigate the fields, rice plants cannot grow properly as they are drown in the muddy water and sometimes, the paddy fields are even drying out. As a result, the yield goes down to 10 baskets per acre according to experience in 2014 (IDGY-28).

According to the perceptions of experts (IDGY-07, 08, 10, and 11) the current and future potentials of farming situation in the area are the following:

– Paddy will be exported to Hopin, Mohnyin, Hpakant and Myitkyina. Added value of products should take into account in the future, when the electricity supply will be available. If medium or large-scale rice mills could be implemented in the area, farmers could export not only paddy, but also rice. Currently, there are only small-scale rice mills and the product quality is low, therefore farmers have to sell raw material and cannot consider further food processing.

– Farmers use very little fertiliser and pesticide and some farmers do not at all. This is a great potential for producing organic food or at least almost organic food, if using fertilizers cannot be avoided completely.

Fishing

Fishing is the third most mentioned income-generating activity of the surveyed households and it is mostly a side job for most of the local people. However, it is the main economic activity not only for seasonal migrants from Sagaing and Mandalay Divisions but also for some migrants from Shan State since 1990. Today, 12 fishing-free zones, where fish reproduce and fishing is prohibited, are established in the lake (more details about the fishing-free zones see subchapter 7.2) (IDGY-46). The rest is governed by the freshwater fishery law (State Law and Order Restoration Council 1991), which was enacted in 1991 for Kachin State (IDGY-43). Since 2008 the fishery office was set up in Loneton village and today it has three staff members (IDGY-43 and 48).

Fishermen need to have a license, issued by the fishery department, for their fishing gears, which costs between 5,000 Kyat (approx. € 3.85) and 6,000 Kyat, (approx. € 4.62) per year. The license costs depend on the quantity (the production levels and capacities of the gears) (IDG-48). There are fishermen from 16 villages around the lake active in fishery (IDG-43). According to the registration the number of fishermen in the study area are 60 in Lonsent, between 35 and 40 in Nammoukkam, 85 in Hepa, between 35 and 40 in Shweletpan, 40 in Mamomkai, 60 in Loneton, 13 in Nanpade, 25 in Nammilaung and 61 in Nyaungbin (IDGY-48). However, it is very rare to find fishermen in Main Naung because it is four km away from the lake (IDGY-32). All in all, fishery is an important income generating activity in the area (keep in mind fishing is an easy side job of all people around the lake without registration) (IDGY-31). According to the questionnaire households, which are engaged in fishing have an annual income of 3.0 Mill. Kyat (approx. € 2342) on average. This is the second highest income after agriculture.

Indawgyi fish are famous in the region and the consumers prefer them. Therefore there is a strong market and brokers from Hopin and Mohnyin collect fish and transport them to Myitkyina and Hpakant (IDGY-08 and18). Over the time fish demand has increased constantly, consequently, fishing has intensified especially by seasonal migrants (IDGY-21 and 25). Therefore, some experts (IDGY-18, 25 and 36) complained that most people especially migrants ignore the closed season (June to August, thus allowing spawning and recruitment (IDGY-43)) and practice fishing throughout the year. Moreover, very small mesh size fishing gear is used sometimes and even illegal fishing methods have been reported (often in Shweletpan village (IDGY-05)) (IDGY-25 and 36). Fishing with small mesh size nets under 0.75 inches is prohibited and fishing practices such as poisoning, chemical impact, use of dynamite or electrofishing are strictly illegal. (IDGY-03). According to experts:

"[…] All depend on the fishing sector, thus closed season makes us difficulty" (IDGY-36).

"[…] Local people obey the fishing method, however, in the closed season they still fish, nevertheless, I can understand them well, during three months they need alternative income, till now there is none"(IDGY-07).

One expert from Mohnyin Degree College also noticed:

"[…] Local people have the awareness of overfishing, it would be very nice, if there is a alternative income generating (like handicraft…) because they need it" (IDGY-49).

According to an expert from Friend of Wildlife (FOW-NGO) a micro-finance programme for fishermen, who are interested in financial support, is carried out by FOW in Mamomkai, Loneton, Leponlay, Nammoukkam, Lonsent and Nanpade. In every village a maximum of 30 fishermen and a minimum of 12 fishermen are provided with the maximum credit amount 50,000 Kyat (approx. € 38.46) for six months. FOW works together with Government Cooperation Department, which monitors the progress.

Experts form Nyaungbin and Lonsent (IDGY-13 and 21) pointed out that the number of fishermen has constantly increased in the village and their income has decreased because the fish density is deteriorating. Therefore, fishermen catch smaller amounts of fish compared to previous times (IDGY-10). However, an expert from the fishery office in Loneton explained that the number of fishermen does not increase significantly. Every year only three or four persons, who are related with former season migrants, immigrate. He also mentioned a social problem between local fishermen and migrants:

"[…] Migrants' fishing method is different than local people do. For example, migrants use spear, which is not prohibited according to freshwater fishery law, to catch the fish, while local people use fishing net. Sometime, local people complain that migrants catch the fish, which is inside their net" (IDGY-48).

The possible negative impacts of gold mining economic activity not only affects agriculture but also fishery. Here one fishery officer from Mohnyin (IDGY-43) complained that Main Naung stream, which is one of the inflowing streams of Indawgyi, is drying out and it sediments into the lake. As a result the considerable content of mercury harms the quality of water and water resources in the lake.

According to experts from fishery department in Mohnyin and Loneton (IDGY-43 and 48) the current and future potentials of fishery in the area are the following:
–   As staff and facilities are not sufficient, fishery staff will work together with police and village community, to protect the offspring during the reproducing time by efficiently patrolling the fishing-free zones. Fauna and Flora International (NGO) will support the charges of patrolling.
–   The education programme will be conducted by fishery staff especially in reproducing time in every village around the lake.
–   The fishery department organizes to breed fish (the domestic species) in the pond and in order to reinforce the density of the fish in the lake in August (after the natural reproducing time).

One expert (IDGY-10) pointed out to consider the added value of fishery products, which creates more income in the future. But in doing so the sufficient electricity supply and knowledge sharing or vocational training are necessary as well (IDGY-21). Currently, for example, fishermen sell raw material to brokers due to lack of cool storage facilities, and lack of investment. In the study area (in Nyaungbin) only one small-scale ice mill is located, but it is only of limited use, because it is operated by a generator.

Retail Shop

Retailing is the fifth most mentioned income-generating activity of the surveyed households. The shops are mainly grocery shops. Goods are imported from Hopin and Mohnyin. Their sale amount/volume per day ranges from 5,000 Kyat to 50,000 Kyat (approx. € 3.85 to € 38.46). Most of the larger retail shops are located in Main Naung and Nyaungbin village. More than 10 retail shops are operating in Main Naung, while only three small retail shops belong to owners in Shweletpan.

Other self-employment

Other self-employment is the sixth most mentioned income-generating activity of the surveyed households. This includes mostly the owners of gold mines. It also consists of rice mill owners, car renters, elephant renters, traders and restaurant owners. Gold mines are found along the lake's inflowing streams, near the villages of Nyaungbin, Nanttaungse, Nammilaung and Nanpade (IDGY-23 and 24). Some illegal gold mines can be found near Mamomkai and Main Naung in the conservation area (IDGY-05).

Gold mining on a commercial scale has developed in Kachin State since 1995 (Hla Hla Than 2006). One year later (1996) this commercial scale has started in the Indawgyi region (IDGY-24). In the past, before 1996, people around the lake region searched for gold in the traditional way and it was a side job (IDGY-28 and 32). According to experts (IDGY-23, 24 and 32) the intensified gold mining activity is booming in the area since 2000 and this business was very profitable between 1996 and 2002. Today, gold mining is still a popular income generating activity in the lake area besides farming and fishery (IDGY-08).

Most of the mining seems to be operated by Chinese companies (Khun Sam 2006), thus, gold mining owners are mostly migrants. One expert (IDGY-25) mentioned that KIA possesses some gold mines, especially in the forest near Main Naung. However, there are some local owners, who operate small and medium scale mining (IDGY-30). The local owners are mostly from Nyaungbin, Main Naung, Mamomkai and Nammilaung. According to experts (IDGY-09, 27 and 30) some farmers especially from the west of the lake (e.g. Mamomkai, Nammilaung, etc.) sold their farmlands and invested in gold mining as the income generated by farming was not sufficient (IDGY-34). Some experts (IDGY-14 and 25) noticed that gold cannot be found in the east of the lake.

In general, two to three years after investing in this business, the income is good (sometimes it takes five to six years, this economic activity totally depends on the luck and hope (IDGY-23)). After that the income decreases and sometimes the owner even cannot pay wages regularly to his staff, because gold cannot be found (IDGY-24). It proves that gold mining activity is an unsustainable income source. In addition, the owner has to pay a monthly tax to Army and KIA (IDGY-23). Nevertheless, most of the owners cannot return into the farming sector even if they wanted to because they are in debt.

Experts (IDGY-23 and 24) explained that mine owner and mining workers camp inside the mining area and normally stay for the whole year, but very few go back home, when they have housework. Their families stay at home and wait for their remittance. The wife of a mine owner said:

> "[…] I am just a housewife and take care of two children, who attend primary school. And I am waiting for husband's remittance. Even our couple live separately almost the whole year, we do not have any social problem."(IDGY-23).

Major health problems of mining workers are malaria disease (medical doctor is not available in the mining area) and accidents, which are mostly caused by landslides due to heavy rain and open pit (IDGY-23). One mine owner, who is an ex-farmer, explained his experiences:

> "[…] Now I am suffering from malaria, which spreads in mining area. Man can only work for mine a maximum of 20 years, because it is hard work. Thus, I will work for it as much as I can. Then, I will reverse to farming. I still own farmland and livestock" (IDGY-23).

In addition, another prevalent social problem is drug abuse in the mining area (IDGY-07). One expert pointed out:

> "[…] In Hpakant, where jade mining is located, the owners use Yaba[16] and they do not directly provide the drug to workers. But their assistants do it to gain the efficient work" (IDGY-24).

---

16   Yaba is a combination of methamphetamine (a powerful and addictive stimulant) and caffeine.

But another expert urged:

> "[…] Most of the gold mining owners, who I met, are normal business men and they do not provide the drug to mining workers" (IDGY-30).

However, some experts (IDGY-07 and 31) complained about the consequences of gold mining activity: Drug consumption has started in the mining area and drugs were only available there in the past. But today, drugs are a hot issue in the whole area and every village has drug users and addicts (mostly between 18 and 30 year old) aside from smugglers and vendors.

Moreover, the other visible threat of gold mining is the blocking of streams or diversions and sedimentation in the lake (between Shweletpan and Mamomkai village) by mud produced by the gold mines. A very dangerous threat, however, is mercury contamination. Mercury is widely used in hydraulic gold mining: when gold is extracted from gravel or crushed rock, it has to be dissolved by mercury (IDGY-23). Then the gold is recovered from the solution and mercury is released into the water and the environment. Therefore, mining negatively impacts the agriculture and fishery sector (mentioned in the above subsection on 'Agriculture' and 'Fishery'), but this activity also creates job opportunities for the local people.

Casual work

Casual work is the second most mentioned income-generating activity of the surveyed households. Here, work in agriculture and gold mining are predominant in the area. Agriculture workers (mostly local people who own neither farms nor boats) are hired especially for paddy cultivation in the rainy season either on a daily or monthly basis. Sometimes, the payment is not done in money but in paddy. After paddy harvesting the workers work either in gold mines near the village or in jade mines in Hpakant. In contrast women from west of the lake mostly work as agriculture workers for beans and vegetable farms in the east of the lake, because they prefer to stay in the village (IDGY-31). If they work in gold mining, it is not possible to commute every day. The wife of one gold mining worker explains:

> "[…] I worked together with my husband in the mine. Since three years ago I came back to the village because I need to take care of my child. Now I work as an agriculture worker in the village" (IDGY-23).

Compared to people from east of the lake people from west of the lake are frequently engaged in casual work in gold mining due to the one cropping system (see above subsection on 'Agriculture'). However, seasonal migrants from the Sagaing and Mandalay Divisions and a few from Rakhine State prefer to work as mine workers, earning daily wages and staying in the mines. The daily wage of casual workers in gold mining is 5,000 Kyat (approx. € 3.85) for male and 4,000 Kyat (approx.. € 3.08) for female due to intensified physical work (IDGY-23). In the area agriculture workers earn about 1,000 Kyat (approx. € 0.77) less than gold

mining workers do. Since the gold mining economy started to boom, mostly young
men are interested to work in gold mining.

Today, a lack of manpower for the agricultural sector exists in the villages
(IDGY-23). But, the shortage of casual workers in agriculture is not only caused by
the attraction in mining. Another important factor is that there are a lot of drug users
in the villages, who are not able to work steadily. One expert explained in general:

> "[...] People prefer to get everyday income and by doing so they can foresee their income
> compared to fishing and farming [depend on weather]" (IDGY-23).

But another expert urged:

> "[...] Only few people could save money. But some could not do and their moral is deteriorated
> due to drug abuse" (IDGY-20).

An expert confirmed:

> "[...] In the mining area drugs are very easy to get and the area is out of authority control,
> moreover mining work is hard and dangerous, some tasks cannot be done by normal workers.
> Only the drug users can finish it" (IDGY-30).

In sum, gold mining can create a job opportunity for unskilled people. But it
provides neither sustainable income nor social well-being.

Government or private staff

Working in government or as private staff is the fourth most mentioned income-
generating activity of the surveyed households. Most of them are government staff
in respect of administration, fishery, conservation, education and health. These
people are quite often not originally locals, but have been migrated to the area from
the outside.

Other activities

Besides the above mentioned economic activities eco-tourism can be one of the
potentials. It is based on the nature of the area (lake, landscape, migratory birds,
fish, ethnic tradition, hiking trail, etc.) and is able to support the conservation goals
of the area (IDGY-08, 10 and11). An expert (IDGY-11) mentioned that there is
almost no tourism until yet and only in Loneton two accommodation facilities (one
belongs to the army, which is in general not available for tourists) do exist and one
family accommodates visitors in homestay as well). Currently (December 2015),
the accommodation fees are between € 6 and € 10 per night for foreigners and the
prices are about € 2 less for locals. According to an expert (IDGY-02), international
visitors – in particular researchers or experts – come from October to March and
the local visitors come from September to March, especially during the annual
pagoda festival, which is held in March. However, the number of visitors for both
categories is still low, as the statistical data from 2013 to 2015 reveal: every year

less than 100 international visitors arrive and not more than 150 of local visitors come to the area (not included the visitors of the pagoda festival). One expert pointed out:

> "[…] Ten floor hotel on the shore of the lake is not necessary, the landscape should not be changed too much, eco-tourism means a smaller number of tourists, who maybe pay more, so there is income for the people, who are living there around the lake. Of course, eco-tourism is something, which has the aim to be sustainable: the tourist has to behave in a sustainable manner, and the people living there also should have sustainable manner, not only the relation with the tourist, but I think it is necessary to have a system"(IDGY-11).

The perceptions of some experts (IDGY-09, 12 and 25) are: a hotel zone in the area is not desired, thus just homestay is to develop in the area. Plus, local people should develop and maintain their culture and tradition. One expert also advised to develop community-based tourism in the area:

> "[…] Villages around the lake agree upon special rules that possibly only local guides, local services and something like a service sharing would be there, otherwise non-locals will take over, out of the region" (IDGY-12).

Furthermore, she stated that unity around the villages is very important. This is a necessary precondition for bringing the people together in order to collaborate and to avoid competition. For example, building a hotel or a restaurant costs a lot of money, therefore one solution is: people have to contribute and share, which can be made by place-based or community-based collaboration. An expert from Mohnyin suggested:

> "[…] All hotels should be in Mohnyin. If the transportation is very good, Indawgyi Lake is not far, only 30 miles (48 km) from Mohnyin, it will take one hour. All infrastructures (hotel, banks) should be upgraded in Mohnyin. The weather is mild in Mohnyin and in summer it is not extremely hot. The landscape is very nice. Indawgyi Area should be conserved as nature" (IDGY-26).

The perception of the township and national level experts are very general which means that everyone can invest and a hotel zone can be built in the development zone (see details of the zoning system in chapter 7) of the area.

Nevertheless, almost all experts agreed that eco-tourism can create job opportunities for the people in the area. But in how far local people can benefit from it, depends on the policy and the amount of support for local initiatives.

From the environmental point of view the following suggestions, made by experts (IDGY-10, 11 and 25), should be taken into consideration:

– implementing an education centre, which gives good insights into the wildlife and into the problems of conservation,
– upgrading the knowledge capacity of local people for eco-tourism,
– using non-motor boats on the lake (until now registering boats does not happen).

One of the future opportunities of eco-tourism could be that out-migrated people are attracted to come back to the area because of new good job opportunities. For example, one expert mentioned:

> "[…] Tourism is definitely a good opportunity for people living abroad, who speak good English and have experience in management, etc."(IDGY-11).

According to some experts (IDGY-07, 12 and 36) until now there is no unique handicraft in the area, which can be attractive for tourists.

6.2   HOUSEHOLD FINANCIAL SITUATION

In the area households can be considered as economic enterprises, which are involved in the above mentioned economic activities. In the questionnaire the households were asked, how much income they generate per year with their activities and how much they invest. The difference between both these values is called the annual net income. Besides that the households were also asked for their expenditures (like cost for food and non-food). Taking the difference between the net income and the expenditure is called the balance. Insofar the net income provides information, how much the household earns as an economic enterprise through its activities. The balance gives information about the money, which a household retains after the living costs have been deducted. In reality some respondents were not willing to explain their income fully. The impression was that the information about the expenditure was given quite correctly. Insofar the balance values might be in some cases lower than they really are, but that does not have much influence on the relation of the balances between the households. The following presentation will focus on income and balance, because the expenditures very much depend on the size of the households.

Income – general

In the study area the household annual net income is at an average of 4.9 Mill. Kyat (approx. € 3,769), ranging from a minimum of 0.2 Mill. Kyat (approx. € 154) and a maximum of 48.7 Mill. Kyat (approx. € 37,462). In the frequency diagram of figure 6.3 the annual net household income is classified into seven categories with a width of 2 Mill. Kyat (approx. € 1,538). Over half of the total interviewed households (59%) earns annually less than 4 Mill. Kyat (approx. € 3,077). Only 12% gain an income, which is higher than 8 Mill. Kyat (approx. € 6,154). The remaining 29% of the households stated to earn a middle income between 4 and 8 Mill. with the tendency to be located in the class 4 to 6 Mill. Kyat (approx. € 3,077 to € 4,615).

Figure 6.3: Frequency diagram of the annual net household income

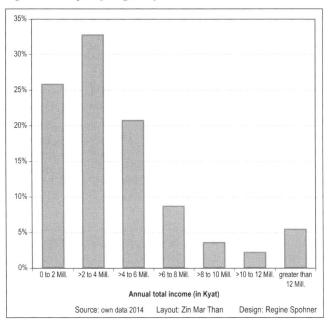

Income – different activities

The boxplot of figure 6.4 and the parameters in table 6.2 show clearly that the different economic activities contribute fairly differently to the income of the households. There are similarities between some activities: 50% of the households (= median) engaged in farming or fishing activities earn more or less the same income below 2.4 Mill. (approx. € 1846) and in the activity 'other self-employment' (mine owner, rice mill owner, etc.) the 'lower' 50% of households get slightly more (up to 2.7 Mill. Kyat (approx. € 2077)). In the retail shop activity the 'lower' 50% of households earn a little less than 2 Mill. Kyat (approx. € 1538). In both the activities, government/private and casual work, households achieve extremely lower income compared to income in the above mentioned activities. In particular for the activity casual work this holds true: the maximum income of 2 Mill. Kyat (approx. 1538) is still lower than the medium income in the activities farming, fishing, retail shop and other self-employment. Besides the lower level of the income in these two activities (government/private and casual work) also the difference between the incomes are not remarkably high (esp. for casual work) as can be seen in the diagram (both the whiskers and box have small lengths). In contrast, the incomes in all other activities show a tendency to higher values. In particular quite a number of outliers do exist.

Table 6.2: Parameters of the net income differentiated by economic activities

	Annual net income from economic activity (in 1000 Kyat)					
	Farming	Fishing	Retail	Self-employ-ment	Govern-ment/pri-vate staff	Casual work
Minimum	201.0	379.0	540.0	250.0	180.0	14.0
Maximum	28,535.0	15,835.7	13,680.0	36,500.0	4,560.0	2,000.0
Arithm. mean	3,663.3	3,045.0	2,629.5	4,940.1	1,248.0	654.8
1. Quartile	1,691.3	1,592.5	1,440.0	1,440.0	700.0	240.0
Median	2,300.0	2,344.2	2,032.5	2,700.0	1,224.0	500,0
3. Quartile	4,557.5	3,661.0	3,360.0	4,800.0	1,296.0	1,080.0
Stand. Dev.	3,907.3	2,545.2	2,342.4	7,432.6	858.3	534.3

Figure 6.4: Boxplot: annual net income differentiated by economic activities

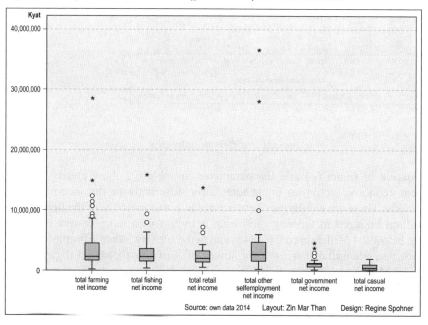

Income – east and west

According to the boxplot of figure 6.5 and the parameters in table 6.3 the household incomes east and west of the lake do not differ very much on average. But really extreme high household incomes of about 30 Mill. Kyat (approx. € 23,077) and above can be found only in the west of the lake with 48.7 Mill. Kyat (approx. € 37,462) as maximum. Households in the east gain a maximum of only 18.5 Mill. Kyat (approx. € 14,231). The boxplot also shows that in general quite a number of outliers do exist in the west, whereas the income differences between the house-holds are not as strong as in the east. Moreover – looking at the box – the range of the box for the west is smaller than for the east, which indicates that the income differences for the middle 50% of the households is bigger in the east. Further, the upper box part in the east is higher than in the west, which means that the income differences of the households in the third quarter are bigger in the east compared to the western counterpart.

Table 6.3: Parameters of the annual net income differentiated by the location

	Income (in 1000 Kyat)	
	Income west	**Income east**
Minimum	201.0	354.0
Maximum	48,720.0	18,479.5
Arithm. mean	4,864.4	4,959.1
1. Quartile	1,806.5	2,230.6
Median	3,336.7	3,649.5
3. Quartile	5,272.5	6,350.4
Stand. Dev.	6,938.8	4,190.9

Figure 6.5: Boxplot: annual net income differentiated by the location

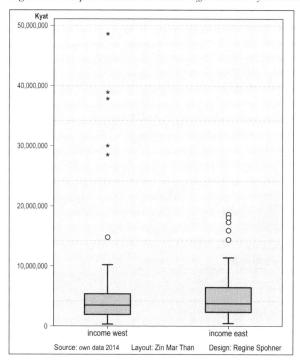

Balance – general

In the research area the annual household balance is at an average of almost 2 Mill. Kyat (approx. € 1538), ranging from a minimum of minus 6.5 Mill. Kyat (approx. € 5000) and a maximum of 44.4 Mill. Kyat (approx. € 34,154). Figure 6.6 shows the frequency distribution of the annual household balance classified into eight categories with a width of 2 Mill. Kyat (approx. € 1538). About 33.3% of the interviewed households show a negative balance. In particular, these households are in the balance class minus 2 Mill. Kyat to 0 Kyat (minus € 1538 to € 0) balance, which means they do not have a very severe negative balance. For more than half of the interviewed households the balance is between 0 and 4 Mill. (€ 0 to € 3077) in particular many households have a slight positive balance ranging from 0 to 2 Mill. Kyat (€ 0 to € 1538). The middle balance classes between 4 and 8 Mill. Kyat (€ 3077 to € 6154) consists of only a little more than 7% of the total interviewed household. A little more than 5% of the total households have an extreme positive balance with more than 10 Mill. Kyat (approx. 7692).

Figure 6.6: Frequency diagram of the annual household balance

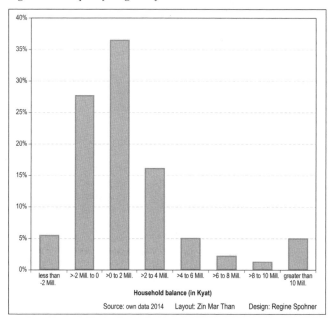

Balance – east and west

The parameters in the table 6.4 show clearly the extreme difference in the household balance between east and west of the lake. While the west ranges from a minimum of minus 6.5 Mill. Kyat (approx. € 5000) and a maximum of 44.4 Mill. Kyat (approx. € 34,154), the east has only a minimum of about minus 3.9 Mill. Kyat (approx. minus € 3000) and a maximum of 15.9 Mill. Kyat (approx. € 12,231). However, looking at the average household balance, the eastern households have about 2.4 Mill. Kyat (approx. € 1846) while the western ones have only 1.7 Mill. Kyat (approx. € 1308).

Table 6.4: Parameters of the annual balance differentiated by the location

	Balance (in 1000 Kyat)	
	west	east
Minimum	-6,533.6	-3,581.5
Maximum	44,379.4	15,965.7
Arithm. mean	1,689.5	2,397.4
1. Quartile	-627.3	58.2
Median	322.6	1,437.7
3. Quartile	1,807.0	3,192.4
Stand. Dev.	6,190.7	3,787.5

Figure 6.7: Boxplot: annual balance differentiated by the location

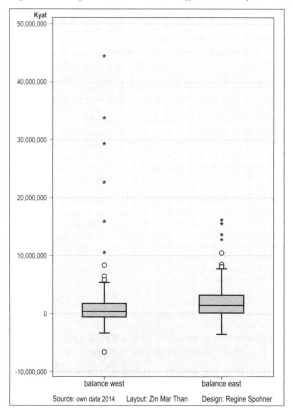

As the boxplot of figure 6.7 points out the differences of household balance in the west are extremely higher than the balances in the east. In particular quite a number of outliers, which shows high values, can be found more often in the west. Moreover, the range of the box of the west is smaller than of the east, which means that the balance differences for the middle 50% of the household is bigger in the east. In the west the box is located around 0, which means that a part of the middle 50% of the households have a negative balance. On the contrary, the box of the east is completely located in the positive interval. It has to be pointed out, too, that the heights of the upper and lower box parts of both, the east and the west, do not differ much, and the height of both boxes are fairly small, which shows that the differences around the medium are not very distinct.

6.3   INCOME ACCORDING TO HOUSEHOLD MEMBERS

In the subchapters above, income and balance are considered according to the households in total without keeping in mind that the household size and the occupation of the members differ. For this reason the focus of this subchapter is on the contribution of annual net income of the economically active members of the households. In order to deal with this topic the income of the activities, each household is involved in, is divided up to the active household members according to their occupation which have been asked for in the questionnaire. Considered is not the income, which is classified as other and which is generated in the households beside the main occupation the members have. If several members are generating household income according to the same occupation, this income is subdivided into equal parts and these parts are assigned to the members. One important question is, whether occupation and educational level of the people do have an influence on the amount of income the members are generating.

Table 6.5: Income of household contributed by household members – total

	Total (in 1000 Kyat)
Minimum	27.0
Maximum	28,000.0
Arithm. mean	1,455.2
1. Quartile	472,5
Median	1,000.0
3. Quartile	1,767.3
Stand. Dev.	2,013.1

Altogether for 554 household members the income, which they contribute to the total annual household income as described above, could be assigned. As table 6.5 shows these household members contribute with an average of about 1.5 Mill. Kyat (approx. € 1154) to the household income. The minimum, a member is contributing, is only 27,000 Kyat (approx. € 20.77), the maximum is 28 Mill. Kyat (approx. € 21,538). 50% of the members contribute less than one Mill. Kyat (approx. € 769).

Table 6.6 also points out, that the incomes are quite different according to the occupation/activity of the members. The lowest average income is generated in casual work, high incomes can be earned when having a retail shop or being engaged in another self-employment activity.

Table 6.6: Household income contributed by household members – activities

	Income from economic activity (in 1000 Kyat)					
	Farming	Fishing	Retail shop	Self-employ-ment	Govern-ment/pri-vate staff	Casual work
Minimum	100.5	402.9	360.0	150.0	180.0	27.0
Maximum	12,319.5	15,835.7	13,680.0	28,000.0	4,560.0	1,610.0
Arithm. mean	1,454.0	1,724.1	2,166.2	2,589.0	1,009.0	405.2
1. Quartile	570.3	862.7	1,065.0	1,000.0	485.0	156.3
Median	886.7	1,172.1	1,440.0	1,800.0	1,080.0	325.5
3. Quartile	1,791.8	2,117.7	2,268.0	3,000.0	1,269.0	500.0
Stand. Dev.	1,746.4	1,909.9	2,187.4	3,930.5	719.1	354.6

Income classes

The frequency distribution of table 6.7 (classes with income spans of 0.5 Mill. Kyat) shows that the vast majority of the family members generate an income up to 1.5 Mill. Kyat. A reasonable number of members does exist in the middle range (from 1.5 Kyat to 2.5 Mill. Kyat). In the classes ranging above 2.5 Mill. Kyat always only a small number of persons is included.

Table 6.7: Frequency table of the income of single household members

Income group (in Kyat)	Abs. frequency	Percentage frequency
0 to 0.5 Mill.	155	28.0
0.5 to 1.0 Mill.	124	22.4
1.0 to 1.5 Mill.	120	21.7
1.5 to 2.0 Mill.	42	7.6
2.0 to 2.5 Mill.	44	7.9
2.5 to 3.0 Mill.	25	4.5
3.0 to 3.5 Mill.	5	0.9
3.5 to 4.0 Mill.	10	1.8
4.0 to 4.5 Mill.	4	0.7
4.5 to 5.0 Mill.	5	0.9
more than 5.0 Mill.	20	3.6
Total	554	100.0

Considering this typical distribution presented in table 6.7 and the conditions for running contingency tables the following five 'personal income' categories have been chosen for the following contingency analyses:
−   0 to 0.5 Mill: very low income generating,
−   0.5 to 1.0 Mill: low income generating,
−   1.0 to 1.5 Mill: medium income generating,
−   1.5 to 2.5 Mill: high income generating,
−   more than 2.5 Mill: very high income generating.

Spatial variation of income of household members

The income of the household members shows a significant difference according to the spatial level of west and east of the village as well. As shown in the table 6.8 incomes west of the lake are very often in the extremely low class and seldom in the low and the medium class; in the east it is the opposite. No significant differences do exist in the high income classes. The contingency coefficient is C = 0.284.

Table 6.8: Income differences per person: 'west' and 'east' of the lake

	Income per household member (in Mill. Kyat)				
	0 to 0.5: ext. low	0.5 to 1.0: low	1.0 to 1.5: medium	1.5 to 2.5: high	> 2.5: very high
West	++++	--	--		
East	----	++	++		

Legend see table 4.2

Table 6.9: Income differences per person on the village level

	Income per household member (in Mill. Kyat)				
	0 to 0.5: ext. low	0.5 to 1.0: low	1.0 to 1.5: medium	1.5 to 2.5: high	> 2.5: very high
West					
Nyaungbin		--			
Nammilaung	++++			--	
Nampade	++++				----
Loneton		--	++		
Mamomkai					
Main Naung	++++		--		
East					
Shweletpan	----				
Hepa		++++	++	----	
Nammoukkan	----			++++	
Lonesant	----				++

Legend see table 4.2

On the village level the contingency coefficient is with $C = 0.470$ even higher. As can be seen in table 6.9 a sharp contrast exist for the extremely low income class. Nammilaung, Nampade and Main Naung have very many members belonging to this class. All three villages are located on the west side of the lake. The opposite is true for the villages Shwelatpan, Nammoukkan and Lonesant, which are located on the east side: they have extremely few members belonging to the very low income class. Moreover, for Nampade the contrast is extraordinarily stark, because here very few members are in the extremely high income class. The only villages, in which the observed frequencies of all income classes are very much similar to the theoretically expected ones (if equally distributed), are Nyaungbin and Mamomkai.

Income and occupation

The contingency analysis shows (see table 6.10) a significant relation between the occupation of the household members and the amount of income they generate. This relation is quite strong as the contingency coefficient of $C = 0.604$ shows. Persons in casual work belong disproportionally often to the extreme low-income category and are almost in no case situated in higher income classes. Almost the opposite takes place with people, who are self-employed. They belong often to the higher income classes. Persons who are employed in the government or in private companies belong quite often to the medium income class and significantly less to higher income classes. For farming, fishing and retail the situation is more complex. Disproportional low numbers of people with such occupations are in the very low-income group. But, whereas people in fishing and retail belong often to the medium or to the extremely high-income class, farmers belong to the low or the high-income class.

Table 6.10: The relation between income and occupation

Income category (in Kyat)	Occupational group					
	Farming	Fishing	Retail	Other self-em-ployed	Govern-ment/ private	Casual work
0 to 0.5 Mill.: very low	--	----	----			++++
0.5 to 1.0 Mill.: low	++++					----
1.0 to 1.5 Mill.: medium	----	++++	++		++++	----
1.5 to 2.5 Mill.: high	++			++	--	----
above 2.5 Mill.: very high		++++	++	++++	--	----

Legend see table 4.2

Income and school level attained

Significant differences can also be stated between income and the school level attained. But, the contingency coefficient of C = 0.255 indicates that in comparison to the relation 'income-occupation' this relation is not as strong. In particular, table 6.11 shows a difference between lower education and higher education: people with lower education, who attained monastic or primary school are disproportionally often members of the low income category and rarely members of the medium income class. On the contrary, members with higher education degrees are often members of the medium income category and more seldom members of the low-income group. Interesting is also that for the high-income categories the education level does not seem to have an influence, which results in the following conclusion: a low educated household member gains more often low income, and a higher educated member gains more often a more satisfying income, but not necessarily really high ones.

Table 6.11: The relation between income and school level

Income category (in Kyat)	School level attained		
	Monastic/primary	Middle	Higher*
0 to 0.5 Mill.: very low			
0.5 to 1.0 Mill.: low	++		--
1.0 to 1.5 Mill.: medium	--		++++
1.5 to 2.5 Mill.: high			
above 2.5 Mill.: very high			

* Higher = high school and university
Legend see table 4.2

7   ASPECTS OF GOVERNING IN THE AREA

This chapter explains the institutional system of the research area. Subchapter 7.1 describes the political/administrative arena and outlines governance issues. Due to the importance of the environmental and conservational aspects in the area, this subject is discussed separately in a second subchapter 7.2.

7.1   THE POLITICAL/ADMINISTRATIVE ARENA

Indawgyi Lake is an ASEAN Heritage site and has recently been included into the tentative list of the World Heritage Sites (Bhandari et al. 2015). Currently, governmental institutions are working in the area, such as departments for village administration, education, health, wildlife conservation, fishery, forestry, police force, military, but also an anti-governmental one, namely Kachin Independence Army (KIA), which is an ethnic armed group. Most of the government agencies have their offices in the village of Loneton, ethnic armed group camps are located in the forest/jungle near Main Naung and Nanpade village (IDGY-05, 24 and 34).

There are also non-governmental institutions: FFI, FOW, Inn Chit Thu Group and several community forestry groups. These NGOs and the wildlife conservation team are actively engaged in the management of the area's environment (IDGY-05).

Moreover, every village has a village elder group, the so-called village committee. It consists of respected representatives of the village like retired teachers, members of youth association or farmers. These people know the situation in the village well and are willing to work voluntarily for village affairs. One of the group members is selected as a head of the village. The selection is done by the township administration committee and the process is as follows according to one expert:

> "[…] There is a village administration law and we selected five persons from each village according to the law, then we familiarize them with administration procedures. After that they are elected by the village committee and one is appointed as a head of the village" (IDGY-44).

Compared to the number of staff working for the conservation or in the military the number of staff in the other government offices is often very small.

Kachin Independence Army (KIA)

The ethnic armed force, Kachin Independence Army (KIA), will be discussed here in more detail because it has been and will be an important factor in developing the area in the future. The Kachin Independence Army was founded in the beginning

of the 1960s and was aiming at independence for the Kachin state. It is the army of
the Kachin Independence Organization (KIO) and Myanmar's second largest non-
state ethnic armed group (Human Right Watch 2012). The Kachin Independence
Army fought against the Myanmar military government over the years, during this
time the Indawgyi Lake Region was an insurgent area.

In 1994, a ceasefire agreement was signed between the Kachin Independence
Organization and Myanmar military government (Min Zaw Oo 2014). The
agreement had impacts on political, economic, territorial and legal issues. For
example, certain areas would be under exclusive Myanmar government or KIO
control, and other areas would be under shared control. This shared controlling
enables the administrative powers of peripheral areas, which were affected by civil
wars and neglected for many years, to bring in humanitarian assistance and foster
development (Human Right Watch 2012: 24). At that time, KIA leadership moved
down from the mountainside and set up a new headquarter at Laiza, which is now
a border town between China and Myanmar, but the KIA was allowed to retain its
arms and controlled the territory (Thant Myint-U 2012). As a result, a new political
economy has emerged leading to a dramatic increase in the exploitation of natural
resources, with both sides – Myanmar and Kachin – using the connections to China
for cross-border business such as jade mining and trade, toll roads, logging (mostly
illegal), and etc. (Levin 2014, Thant Myint-U 2012: 97).

In the period from 2006 to 2010, the situation began to worsen between the
Myanmar government and Kachin Independence Organization, because the key
person from the government side, the then Prime Minister and Chief of the Military
Intelligence (MI) Gen. Khin Nyunt, who negotiated the ceasefire deals with KIO,
was removed from power in October 2004 (Bünte 2011, Taylor 2012). The
newcomers needed time to build new trust between both parties again. However,
their endeavour to build trustful relations between both parties did not work as well
as before even though they attempted (Taylor 2012, Min Zaw Oo 2014). Another
reason was, that in preparation for the 2010 transition the regime had announced in
2008, that the ethnic armed groups, which agreed to the ceasefire agreement, have
to transform either into a Border Guard Force (BGF) to become a direct part of the
Myanmar Army or into a people's militia unit under the control of the Myanmar
Army (Min Zaw Oo 2014, Bünte and Dosch 2015). But the Kachin Independence
Organization refused a transformation into either Border Guard Force or people's
militia without a political solution for the underlying causes of ethnic tension and
conflict (Nakanishi 2013). As a result, KIO returned to open conflict with the
Myanmar Government. Therefore, in 2009 the Kachin Independence Army began
a voluntary recruitment programme to increase its standing army (Human Right
Watch 2012: 26,). Around 2009 and 2010 both parties had a series of minor
incidents again with a slightly changed and more moderate aim: the Kachin
Independence Organization only wants a federal system of government, with equal
rights for all citizens and a degree of local autonomy (Thant Myint-U 2012: 105).
The 2008 constitution considered such aspects: for instance, the provisions for
ethnic nationality rights, equality and a form of federalism with decentralization
and democratization of power are stated there (Burma News International 2013:

26). However, at the same time revenue sharing and division of power are defined in this constitution only in general terms, without mentioned detailed implementation procedures.

From June 2011 to December 2012 major fighting took place between the two parties, in particular evidenced by the hydropower Tarpine Dam[17] dispute (Human Right Watch 2012: 30, Lorin 2014). However, there were several other grievances, which caused the Kachin Independence Army to end the truce with the Myanmar government, such as: the 2008 constitution (above mentioned), the rejection of KIO leaders' registration as independent candidates in November 2010 for national election, the refusal to allow KIO to form a political party and so on (Human Right Watch 2012: 26, Taylor 2012). In more detail an expert confirmed:

> "KIO is an armed group. So, it is not allowed to be elected as a political party and we never demanded it. But one thing: Dr. Tu Jar, who was the vice president of KIO, left the KIO, and set up a political party. But party registration was not granted in time for election 2010 and 2012 (by-election). In 2015 his party could be elected but has lost" (IDGY-38).

During the first parliament period (2011–2016), exactly in August 2011, the government invited ethnic armed groups in the country to initiate a peace process with all armed groups aiming at lasting peace (Min Zaw Oo 2014, Taylor 2012, Bünte and Dosch 2015). Therefore, the government assigned two teams (A and B) to negotiate with ethnic armed groups. These teams consist of not only the government officers but also include some representatives from non-governmental organizations. This constellation has not appeared before. In general, team A contacted the groups (e.g. KIO/KIA, UWSA (United Wa State Army), etc.), which signed the ceasefire agreement with the military regime (1994) and team B focused on the groups that did not sign a ceasefire agreement previously (e.g. KNU (Karen National Union) etc.) (Min Zaw Oo 2014). Moreover, an expert of KIO (IDGY-38) explained that the discussion is accompanied by international observers (one from China and one from the UN). This demonstrates some transparency in the current negotiation process, which never was the case before. As a result, in October 2015 the government of the first parliament signed the final version of a Nationwide Ceasefire Agreement (NCA) with eight ethnic organizations including KNU (Wilson 2015). However, only half of the recognized ethnic groups signed the agreement and the KIA, the Shan State Army and UWSA are still absent to sign (Wilson 2015). Therefore, the process is incomplete, and the NCA can unfortunately not bring fighting to an end.

There were hostilities between the government of the first parliament and the KIO all the time. Trust could not been built between the two parties. The government of the first parliament has not achieved a ceasefire agreement with Kachin after several rounds of negotiation, which always seems to end up in restarting the hostilities (Bünte and Dosch 2015). One expert from KIO pointed out:

---

17 The Tarpine Dam was constructed by Chinese company and Myanmar government in KIO territory without first consulting or seeking approval from the KIO or local communities (Human Right Watch 2012: 30).

"[…] On the one hand the government of the first parliament is discussing peace with the ethnic armed forces and on the other hand the military conducted the air strike in Kachin (near Mohnyin in November 2015) and Shan. It shows that the government's wish to negotiate peace with ethnic armed groups is not serious, even though we are approaching discussion about peace optimistically." (IDGY-38).

He agreed that the peace process should be implemented according to the perception of the rule of law, negotiation and constructive engagement. However, he felt that the current peace process cannot be approached only by the rule of law and that it does not involve all the stakeholders. He further mentioned:

"[…] Every ethnic armed group has a policy and principle agreements but in reality the discussion is approached only tactically and strategically and principle agreements cannot be implemented".

"[…] Therefore, we expect the change and just voted for NLD (National League for Democracy). At present we are just waiting for new government (NLD) with our full hope that we will discuss about our aim, which is a real federal union according to the Panglong agreement[18]" (IDGY-38).

How the peace process will continue depends on the offer of the new government (in office since March 2016) and the following talks between KIO/KIA and the new government.

7.2 THE CONSERVATION ARENA

The 2008 Constitution of Myanmar states in Chapter 1 Article 45: "The Union shall protect and conserve its natural environment". It is one of the basic principles of the Union. Additionally, the Constitution states in Chapter 8 Article 390: "Every citizen has the duty to assist the Union in carrying out the following matters: (a) preservation and safeguarding of cultural heritage; (b) environmental conservation; (c) striving for development of human resources; (d) protection and preservation of public property" (Union of Myanmar 2008). The stipulation clearly shows that protecting the environment is not only the duty of the government but also of the citizens. However, there is no specific law, which applies directly to conserve wetland. Conservation of national wetlands is covered by several national legislations such as the Protection of Wildlife and Conservation of Natural Areas Law of 1994 (State Law and Order Restoration Council 1994b) and the Environmental Conservation Law of 2012 (Republic of the Union of Myanmar 2012).

    Since 1999, the Indawgyi Lake is declared a conservation area ('Indawgyi Wetland Bird Sanctuary'), managed by dedicated personnel from the Nature and Wildlife Conservation Division of the Forest Department (UNESCO World Heritage Centre 2014). This organization has emerged according to The Protection of

---

18   The Panglong agreement is an agreement, which has been signed in Panglong between the Burmese government under General Aung San and the Shan, Kachin, and Chin in 1947. The agreement accepts "[…] Full autonomy in internal administration for the Frontier Areas" (Tinker 1984: 404–405, Ethnic National Council of Burma 1947).

Wildlife and Conservation of Natural Areas Law of 1994 (Zin Mar Than 2011). The boundaries of the conservation area can be seen in figure 7.1. In February 2016 Indawgyi Lake has been designated a Ramsar Site (Ramsar 2016) and has been included in the tentative list of natural world heritage sites of UNESCO (UNESCO World Heritage Centre 2014). Moreover, an expert (IDGY-50) informed that the government wants to apply for admission of the Indawgyi Lake into the Man and the Biosphere Programme of UNESCO[19].

Figure 7.1: The conservation area of Indawgyi Lake

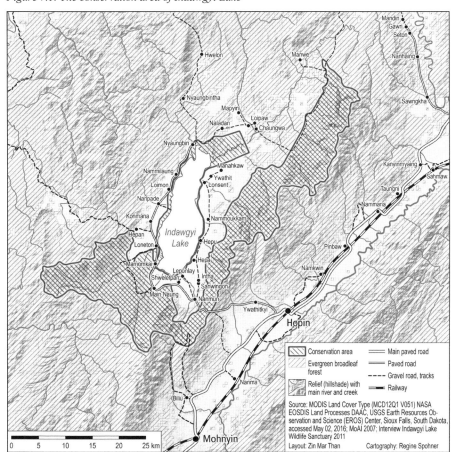

---

19   UNESCO's Man and the Biosphere Programme (MAB) was launched in 1971 and is an Intergovernmental Scientific Programme that aims to establish a scientific basis for the improvement of relationships between people and their environments (UNESCO 2016).

In 2015, the conservation staff in Indawgyi Wildlife account for 14 people (10 are permanent staff and four are daily wages workers) (IDGY-05). Therefore, experts claimed (IDGY-05 and 25) in particular four issues: one is an insufficient number of personnel, the second and third are dealing with an ineffective patrolling, namely insufficient equipment, and the unstable political situation, and the fourth is related to an ineffective and unclear notification of the conservation boundary area as outlined in the following statements:

> "[…] Gold mining near Nyaungbin is in the watershed area, but the area is not conserved" (IDGY-25).

A similar problem is that

> "Forests are exploited for firewood in the watershed area, but it is out of control because the area is not protected." (IDGY-05).

And an expert explained,

> "Our wildlife conservation department is practicing the traditional way, i.e. we do not have zone system according to the specification of IUCN[20] (International Union for Conservation of Nature)" (IDGY-25).

As a consequence, a conflict frequently occurs between conservation staff and local people. According to experts, at the beginning the conservation staff practiced punishment/fine culture, then they started to realize that this does not work well for achieving the goal. One international expert (IDGY-25) also commented that if everything is protected nobody will respect it. But if there is a sequence of zones in accord with the IUCN specifications, then people are more likely to respect that.

Although a general zoning system does not exist currently (in 2015) for the whole conservation area, a system only for the lake is in effect, i.e. there are nine fishing-free zones with a minimum area of 1.3 km^2 and maximum area of 2.6 km^2 each in the lake (IDGY-46). Moreover, it is also forbidden to catch fish within a certain distance from the pagoda in the lake due to religious taboo (IDGY-48). Additionally, two zones near Hepu and Hepa exist to protect specific species. Here it is forbidden to fish within the first 20 m from the bank of the lake (IDGY-46). Apart from these zones small-scale capture fishery is practiced and accepted in the lake and is governed by the freshwater fishery law (already mentioned in subchapter 6.1 section 'Fishing').

Nevertheless, at present the conservation team and FFI (Flora and Fauna International) plan to identify the zones in the area with the involvement of village community groups (IDGY-47 and 50). It will include three categories: first, the core zone, which is only of use for research and totally protected, second, the buffer zone, which is partially protected and partially allows for wise use and third, the development zone, which will be developed for social and economic use (IDGY-47 and 50). The development zone actually consists of farmlands and other economic activity places (e.g. restaurants, shops, guest houses and so on) (IDGY-

---

20 Specification of IUCN means a sequence of zones with different categories, for instance one zone: no use at all, then another zone: slightly used, the next zone: can be used (IDGY-25).

47). An expert (IDGY-47) explained that the zoning system is underway. After it has been implemented, the conservation team will also support education programmes related to the usage of fertiliser and pesticides in the development zone to reduce the negative impact on the lake (e.g. eutrophication). Moreover, he expected that conservation awareness of the local people can be improved and that less conflict between them and the conservation team will arise.

According to an expert (IDGY-48) fishery staff from Loneton will collaborate with the police and village community groups to patrol the fishing-free zone every week. All costs will be covered by the partner organization (FFI). Currently (in 2014 and 2015), the conservation team and a partner organization (Friends of Wildlife (FOW)) organize controls of the fishing practices and patrols of the fishing-free zones in the lake. This partner organization covers all costs including boat for lake patrolling. The patrolling team consists of two rangers, two foresters, one day labourer and four community members and the team patrols about 12 times per month according to upcoming needs (IDGY-05). At the same time some governmental and non-governmental organizations (Flora and Fauna International, Deutsche Gesellschaft für International Zusammenarbeit, European Union and so on) cover all costs including expenses for motorbikes for land patrolling. This team is made up of three rangers, three foresters, four day labourers, a forest guide and two community members or additional rangers (IDGY-05). They patrol once per month, but it takes 10 days. If the patrolling teams find a problematic situation like electrofishing or fishing net in the fishing-free zone or illegal firewood collection or illegal timber exploitation, a two-step procedure of action is taken: in the first step they do not take the action, just give the instructions and warnings to the people who are involved. In the second step, repeated offenders will be fined. A fine could be for example confiscating the fishing gear.

An interesting, but negative finding is, that one fishery officer (IDGY-43) says that the fishery department does not cooperate with wildlife conservation department, which is also responsible for protection.

According to experts, there are serious patrolling shortcomings due to area security issue/unstable political situation of the area. For instance, the land patrolling team informs the head of the village two or three days ahead about their patrol, because the Kachin Independence Army (KIA) is staying in the forest area where the patrolling route runs along. Sometimes, depending on the political situation, KIA does not allow the team to pass through the forest, and then the team cannot work. Under these circumstances one village head complained:

> "[…] The situation of our villagers is very difficult, when the police comes to our village or the patrolling team is around the village, we have to inform the KIA in order that the conflicting groups do not meet each other" (IDGY-24).

Another expert also explained:

> "[…] Sometime the army complains to us that we supported rations to KIA, but we do because the army is situated on the hill and it is well guarded, we all villagers are not and therefore afraid" (IDGY-21).

As a consequence quite a number of illegal gold mines are located in the conservation area near Main Naung village. One expert mentioned:

> "[…] Near Main Naung village there are about 100 gold mining sites, 75 sites of which are located in the conservation area and the other 25 sites are in the vicinity. Near Mamomkai village 25 sites are also located in the conservation area" (IDGY-05).

According to experts (IDGY-05 and 25) these mines are organized by the Kachin Independence Army, and some local people support it. The visible threat of gold mining is the blocking of the stream (Nant Yin Kha) near Main Naung, or diversion by mud produced from the gold mines (see subchapter 6.1). The invisible but most dangerous potential threat however is mercury contamination. The wardens claimed that they are unable to prevent the establishment of gold mines in the preserved area, due to the political situation. Furthermore, the Myanmar 'Mines Law' of 1994 (State Law and Order Restoration Council 1994a) is vague and incoherent. It consists largely of general statements lacking the clarity regarding waste disposal from the mines.

In relation to pollution, Myanmar had no specific laws to govern air and water pollution for a long time. These issues were loosely covered by the Public Health Law of 1972, empowered by the Ministry of Health, however these issues are covered in the natural area by the Protection of Wildlife and Conservation of Natural Area Law of 1994, executed by the Ministry of Forestry. Nevertheless, in 2012 the Environmental Conservation Law is enacted and this law governs all kind of pollutions and waste disposal. Therefore, it can be hoped that in the near future the situation will become considerably better.

In the area the planting of community forest can be observed, i.e. some villagers collaborate to plant trees on vacant land. Quite often bamboo and eucalyptus are planted to use them for fencing and firewood respectively. Of course, from the conservation perspective this might not be a perfect solution for reforestation, but it helps to protect the existing forest. An expert from FOW (IDGY-46) mentioned that they assist community forest in particular through providing nursery plants and helping to apply for the land from the government. The personal observations of the author is that the government (in 2015) is quite sceptical of this approach, the reasons for that seem to be neither environmental nor conservational aspects but more political ones.

A definitely positive aspect for conservation and environmental protection is that environmental education classes are initiated by FOW at the school in some villages (IDGY-46). On the contrary, a lack of knowledge about waste disposal is critical as an expert stated:

> "[…] The local people have lack of knowledge about waste disposal, so it is disposed into the lake there is getting dump especially from gold mining" (IDGY-50).

In general, solid waste disposal is a critical issue to the lake and the waste mostly consists of plastic litter. There is no municipal waste collection. Some people who live in the lake fringe area deposit their household waste at the bank of the lake.

Other people dump the waste outside of the villages (already explained in subchapter 4.2), but these dump sites are flooded in the raining season. One expert explained:

> "[…] Another aggravating issue increasing the waste amount is connected to the annual pagoda festival, which takes place between the end of February and beginning of March for seven to ten days. Everyday about 80,000 pilgrims attend the festival. Most of the pilgrims and all shopkeepers camp on the bank of the lake near the pagoda. All their waste is left on the bank of the lake" (IDGY-09).

The resulting water pollution and – as a consequence of this – the threat to aquatic life depend on the content of household waste and dissolved materials. In addition, the issue of litter in the lake is exacerbated by fishermen, using plastic bottles as fishing tools to float their nets. Plastic waste also blocks soil enrichment and causes soil contamination as well. She further informed:

> "[…] In Nyaungbin, where are the most migrants, the bank of the lake is heavily polluted with waste. A local person explained that his cow died, in its inside a lot of plastic was found" (IDGY-09).

8 EVALUATION OF THE SITUATION BY THE RESPONDENTS

This chapter presents the results of the evaluations of the social and economic situation of households and villages according to the information gained through the household survey and by the expert interviews. It helps to understand more in-depth the needs and perceptions of the locals. Subchapter 8.1 describes the evaluation of the current and future situation in general. Subchapter 8.2 shows in more detail the evaluation of the current and future situation by comparing the villages east and west of the lake. Subchapter 8.3 deals with the ideas households have regarding their future economic activities.

8.1 CURRENT AND FUTURE SITUATION – GENERAL

Current situation – general

The area has difficulties in developing more rapidly due to the unstable political situation and the poor infrastructure such as information and communication facilities, transportation, and electricity supply (IDGY-09, 10 and 11). The data from the household survey also shows that many households acknowledge the poor infra-structure. More than 80% negatively assessed the current electricity supply system while about 60% of them negatively evaluated the information, communication and transportation facilities (cf. figure 8.1). Moreover, 70% of respondents were not satisfied with farm water availability. In addition, almost 90% of the respondents were unsatisfied with the waste management system. Obviously, this result substantiates the situation mentioned in subchapter 7.2. In contrast, only water quality was positively evaluated by more than 70% of the households. Regarding the social aspects, 75% and about 65% of the respondents critically viewed the current health and education facilities respectively.

More than 60% of the respondents have a critical impression of their current job opportunity and the income generated by the households, which is in accordance with the claims of all experts:

> "[…] A lack of job opportunities for educated persons does exist (e.g. in Nammoukkam and Hepa more educated persons live there than jobs are available, the danger of the brain drain is apparent and takes place already)" (IDGY-05).

The poor job situation is combined with a negatively evaluated market accessibility for products (more than 50% of respondents see this aspect critically). Also the development of tourism is currently evaluated not very positive (about 60% of respondents have a negative impression).

More than half of the households (55%) are of the opinion that the conservation area has an adverse effect on the local economy. Going a step further, a lack of environmental awareness in the research area can be reported according to experts (e.g. IDGY-50). These results show a conflict between local people and conservation issues, which urgently has to be solved in order to achieve a proper sustainable future development in the lake area.

Figure 8.1: Current situation – general

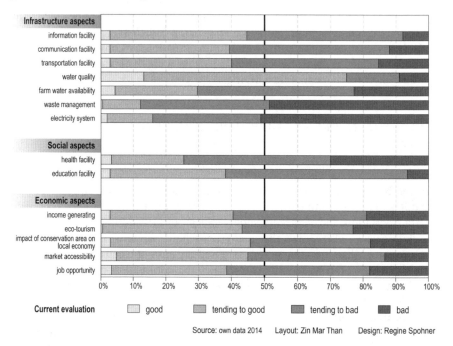

Source: own data 2014   Layout: Zin Mar Than   Design: Regine Spohner

Future situation – general

Figure 8.2 shows the evaluation of the future situation of the area and demonstrates clearly that in general the households see the future more positive. Especially with regard to infrastructure supply, about 80% of the households evaluated aspects like transportation, communication and information technology and electricity supply positively. In addition, economic aspects such as future market accessibility, income generating and job opportunities in the area were also assessed well by more or less 70% of total respondents. However, future eco-tourism development of the area was comparatively critically evaluated by almost 40%. At the same time about 45% of the respondents believed that in the future intensifying the conservation area could have negative impacts on the local economy.

In terms of social aspects, future supply with health facilities in the area was seen positively by 50% of the households and future supply with education facilities was even better evaluated (70% of the households saw it positively). Only one critical issue has to be mentioned: waste management is viewed also for the future quite negatively by 70% of the households.

Figure 8.2: Future situation – general

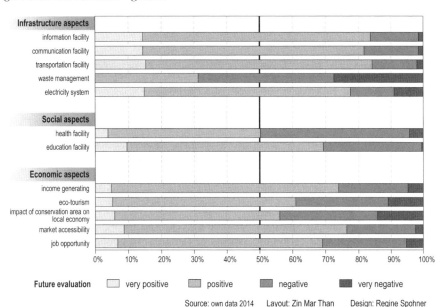

Source: own data 2014   Layout: Zin Mar Than   Design: Regine Spohner

Future household situation

The evaluations of the future household situation related to social and economic aspects are shown in figure 8.3. Regarding to social aspects the households were asked how they see their future hygienic condition and housing condition. These questions – at least partly – give an insight into the environmental awareness and waste management practices of the household. Fortunately, respondents assessed both aspects positively: about 80% do that for the hygienic condition and 75% for the housing conditions.

In terms of economic aspects, household expense, household income generating activities and impact of the conservation area on the household economy were recorded. More than 70% of the respondents saw their future household expenses critically. The reasons for their fear of decreasing household income were diverse. Quite frequently respondents mentioned that school costs, increasing household members and inflation could strain their income in the future. Additionally about 50% of the respondents estimated that the establishment of a conservation area could impact their household economy negatively. In contrast, the future household income generating activities were evaluated optimistically by about 70% of the respondents.

Figure 8.3: Future household situation

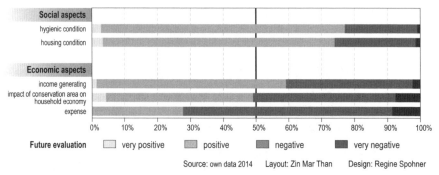

Source: own data 2014   Layout: Zin Mar Than   Design: Regine Spohner

Comparing current and future – general

Figure 8.4 shows the comparison between current and future evaluation of infra-structure and social and economic aspects of the area. In general, the respondents assess the future prospects of the infrastructure situation in the area as well as their household situation positively.

In particular, the households expected the infrastructure aspects such as infor-mation and communication technology, transportation facilities and electricity supply to become much more sophisticated in the future. Often the positive answers have been at least twice as many as for the current situations (for example, the most obvious one is electricity supply: currently 15% of the households had a positive evaluation, 75% of the respondents see a good perspective for the future). The only exception is waste management. Respondents assessed the future prospects also more positive but much less than for all other aspects.

Similar results can be found for social aspects like health and education facili-ties. For instance, the current health facility was positively evaluated by 25% of the households and 50% had positive expectations for the future, quite parallel is the evaluation of education facilities with 35% for the present and 70% for the future.

Regarding the economic aspects, no tremendous differences in the evaluation between the present and the future were reported. Although a positive trend in the evaluation can be stated for all aspects, in some cases such as in eco-tourism and impact of conservation area on the local economy only 10% and 15% of the respondents saw a shift into a positive direction.

Figure 8.4: Comparing current and future – general

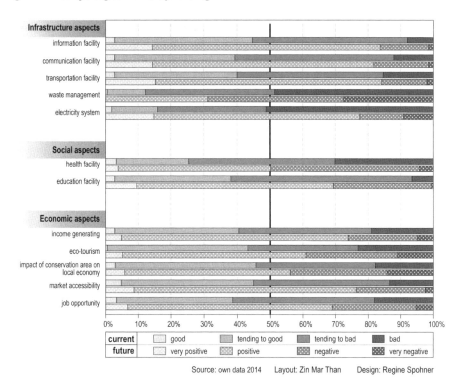

Source: own data 2014   Layout: Zin Mar Than   Design: Regine Spohner

8.2   COMPARING THE EVALUATION OF CURRENT AND FUTURE SITUATION OF EAST AND WEST

As has been pointed out already in chapter 6 differences regarding the economic activities exist between the villages east and west of the lake. In this subchapter it will be investigated if such differences also can be found with respect to the respondents' evaluation of their economic situation and the infrastructural situation they find themselves in.

Current evaluation – East and West

The results of the current evaluation of infrastructure and social and economic aspects divided up in east and west villages is shown in figure 8.5. In general, the respondents from the east evaluated most of the aspects more critically than the respondents from the west. This difference is at first of some surprise because the economic situation (e.g. household income) is better in the east as has been described in chapter 6. However, the result is understandable when keeping in mind one important fact: the road in the east has been upgraded to become a gravel road while only the road in the west has changed into a paved road (see subchapter 5.1). In general, the differences in the evaluations of these aspects for east and west varied between a maximum of 20% and a minimum of 5%.

Of course, a few exceptional cases also could be found. For example, the infrastructure aspect 'farm water availability' was more critically evaluated by the respondents from the west. It is not a surprise keeping in mind the statements of the experts which have been mentioned in subchapter 6.1 (subsection 'Agriculture'). In addition, the evaluation of water quality, waste management and electricity supply were more or less the same for both sides. Moreover, concerning the economic aspects both the sides (east and west) had a very similar perception regarding the negative impact of the conservation issue.

Figure 8.5: Current evaluation – east and west

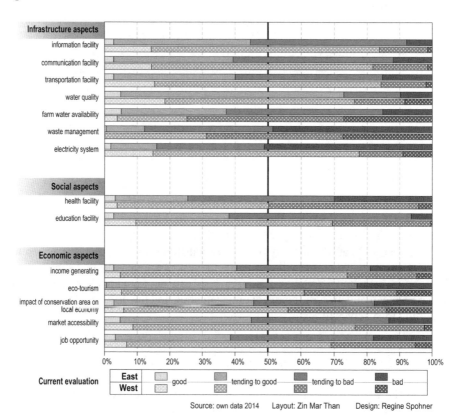

Source: own data 2014   Layout: Zin Mar Than   Design: Regine Spohner

Future evaluation – east and west

Figure 8.6 shows the comparison between the future evaluations of east and west. For the future both sides (east and west) had almost the same positive evaluations regarding to infrastructure aspects (with about 80% evaluating it positively). Only electricity supply was seen more critically by the respondents in the east than by respondents in the west. A reason for that might be that some villages in the west have already access to electricity supply while the electricity supply system is under construction in the villages in the east (see subchapter 5.3).

Figure 8.6: Future evaluation – east and west

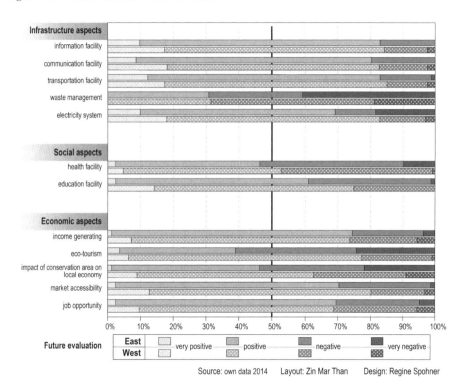

Source: own data 2014   Layout: Zin Mar Than   Design: Regine Spohner

Concerning social aspects, health and education facilities were evaluated similarly, i.e. the evaluations of the household in the east were more critical (differences are maximum 15% or minimum 5%) than the evaluations of their counterparts in the west. With regard to economic aspects both sides (east and west) were very much similar in the evaluation of future income generating potential and future job opportunities. 75% and 70% of the households respectively saw these aspects positively. As seen before, the respondents in the west saw the other aspects such as future eco-tourism development, impact of the conservation area on the local economic activities and market accessibility more positively than by the respondents in the east. The biggest difference could be found in eco-tourism development, which is perceived much higher west of the lake. The reasons might be that at the time of the research accommodation facilities for eco-tourists were located exclusively in Loneton in the west which brought tourists mainly to the western areas. There was also a young volunteer group in Loneton, named Inn Chit Thu, which offered leisure activities to visitors, such as renting of kayaks and bikes. Insofar it seems that households west of the lake had better insights into the kinds of opportunities offered by eco-tourism.

Future household evaluation – east and west

Figure 8.7 shows the comparison of the respondents' perception of the future household situation between the western and eastern lake villages. For social aspects the households on both sides (east and west) evaluated very much similar and quite positive. More than 70% of respondents from both sides positively assessed their future hygienic and housing condition. In detail, the future household hygienic condition is positively rated in both areas by 78% of the households. However, regarding housing condition the respondents from east of the lake were a bit more pessimistic with 28% of the respondents expecting the future conditions to be more negative, while only 25% of the respondents from west of the lake did so.

Figure 8.7: Future household evaluation – east and west

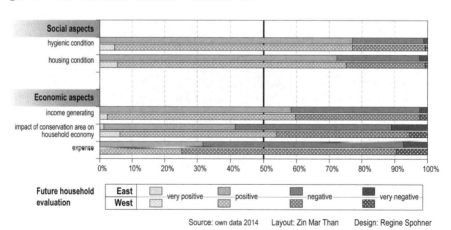

Source: own data 2014   Layout: Zin Mar Than   Design: Regine Spohner

Concerning the economic aspects, the future household income generating potential is very positively evaluated in both areas by almost 60% of the households. Normally, the respondents from the west estimated more optimistically than respondents from the east did. For example, 45% of respondents from the west thought that the conservation area could have adverse effects on their household economy, while 58% of respondents from the east did. This tendency is not true for the household expenses, while 75% of the respondents from the west saw them pessimistically only 68% of respondents from the east did so.

8.3   FUTURE ACTIVITIES OF THE HOUSEHOLDS

Based on the results of the evaluations presented in the previous subchapters, which have shown that the households are quite critical about many of the aspects related to the economic and social living conditions and the infrastructural supply in their areas, this subchapter is dealing with the ideas the households have in carrying out economic activities in the future in order to improve their conditions. The topic is also related to chapter 6 in which the current economic activities of the households have been presented and it is an important information base for setting up endogenous development paths, which will be discussed in Chapter 10.

According to the questionnaire 56.5% of the surveyed households answered that they did not need to change their economic activities. However, 42.1% of the surveyed households wanted to change their economic activities.

Table 8.1: Kind of changed economic activities

Economic activities	Frequency	% of households
Shop keeping	21	30.9
Extend the current business	17	25.0
Jewel trade	10	14.7
Service industry in general	6	8.8
Mining labour	6	8.8
Farming	4	5.9
No answer	7	10.3
Gold mining	5	7.4
Trade	2	2.9
Don't know	3	4.4
Fishing	1	1.5
Other	10	14.7

The households that wanted a change can be divided up into two types:
1.   33.3% (72 households) of those would keep the present economic base and either extend their economic activities or seek to add a new one. A quarter of these households would prefer to extend their present economic activities. In the latter context, economic activities such as shop keeping or jewel trade were quite frequently mentioned by households as possible secondary income sources (cf. table 8.1). Other economic activities such as service industry in general (e.g. car rental, rice mill and ice mill), mining labour, and gold mining were quite moderately often named as new economic activities. An interesting result was also that farming and fishing – the two most important activities in

the area until yet – were of minor interest to be picked up as new economic activities.

2. The remaining 8.8 % would like to give up the present economic base and wanted to establish a new one (cf. table 8.2). However, quite many – more than one third – did not know what kind of new economic activities they should add as income source. Besides that farming, trade and shop keeping were quite frequently mentioned as desired new economic activities, other activities like service in the tourist sector and jewel trade were of less interest.

Table 8.2: Kind of new economic activities

New economic activities	Frequency	% of households
Don't know	7	38.9
Trade	6	33.3
Farming	5	27.8
Shop keeping	5	27.8
Service in the tourist sector	2	11.1
No answer	2	11.1
Service industry in general	1	5.6
Jewel trade	1	5.6
Other	1	5.6

In addition, the reasons for giving up the present economic base were investigated. Two reasons were quite frequently given, namely, 'income is not good enough' and 'young people have other ideas and interests' (cf. table 8.3).

Table 8.3: Reasons for giving up

Reason	Frequency	% of households
Income is not good enough	7	46.7
Young people have other ideas and interests	4	26.7
The work is too hard	1	6.7
No answer	1	6.7
Other	2	13.3

Regarding eco-tourism development in the area, the households were asked whether they were interested to engage in the eco-tourism sector or not. A first important result was that 61.1% of the surveyed households said that they are not interested in eco-tourism and only 35.2% answered that they saw eco-tourism as an additional income source (cf. table 8.4).

Going a step further, the households were asked how they wanted to be involved into eco-tourism. Some households gave more than one answer (cf. table 8.5). Many mentioned that they wanted to open a souvenir shop, offer boat rental, or give guidance to tourists. Some also showed their interest in opening a restaurant or running a home stay. Again others thought about renting cars, opening guest houses or becoming hotel staff. However, most of the respondents – more than half – did not know how to contribute to eco-tourism, although they were interested.

Table 8.4: Interest in eco-tourism sector

	Frequency	% of households
Yes	76	35.2
No	132	61.1
Don't know	1	0.5
No answer	7	3.2

Table 8.5: Involvement in eco-tourism

Activity	Frequency	% of households
Don't know	36	51.4
Gift shop	15	21.4
Boat rental	11	15.7
Guide	8	11.4
Home stay	4	5.7
Restaurant	5	7.1
Motorbike rental	3	4.3
Other	19	27.1

Table 8.6: Important factors having to be improved in the next years

Factor/item	Fre-quency	% of house-holds	Factor/item	Fre-quency	% of house-holds
Health	122	61.6	Security	4	2.0
Education	121	61.1	Farming	4	2.0
Electricity	107	54.0	Water quality	3	1.5
Transportation	90	45.5	Waste management	3	1.5
Communication	16	8.1	Market accessibility	2	1.0
Job opportunity	15	7.6	Environm. awareness	2	1.0
Unity	12	6.1	Don't know	2	1.0
Farm water availability	11	5.6	Technology	1	0.5
Peace	6	3.0	Gold mining free zone	1	0.5
			Other	38	19.2

Finally, the households were asked for the factors (maximum 3) which – they believed – were the most important ones having to be improved in the next years in their villages. 170 households gave three answers, 23 households answered two items and four households mentioned one factor. The results are shown in table 8.6. At the first glance, health and education were quite frequently mentioned by almost

two third of the households. Electricity and transportation were given second priority (about half of the households mentioned these items). Communication, job
opportunity, unity and farm water availability were also quite moderately
mentioned for improvement.

9   FINAL DISCUSSION – CHALLENGES FOR THE INDAWGYI LAKE AREA

This chapter presents an in-depth analysis of the challenges for the Indawgyi Lake Area in respect to its socio-economic development. In the context of the endogenous development approach, which has been described in chapter 2, a discussion of the results presented in chapter 4 to 8 will be done. As has been mentioned in section 2.1.1 Vazquez-Barquero (1999 and 2003) sees four pillars of a regional development strategy, namely the improvement of the competitiveness of local firms, the attraction of inward investment, the enhancement of human capital or labour skills and the building of infrastructure. Based on this context the discussion will focus on the following five topics:
1. Natural resources
2. Human capital and socio-demographic situation
3. Institution or organization
4. Infrastructure
5. Economic structure

9.1   NATURAL RESOURCES

The Indawgyi Wildlife Sanctuary is a Ramsar Site and the most important bird area of Myanmar. The lake and the surrounding with the water, the soil and the climate offer good chances for traditional economic activities like fishing, and farming. Besides that the mineral resource, gold, is an economic base. The extraction has been increased in particular for about 20 years. Although fishing and farming are quite intensive, a wide variety of the endogenous flora and fauna resources remain in the area, which is the reason for the designation as a wildlife sanctuary. The natural beauty of the lake obviously stimulates the potential for tourism. This indeed can create a fourth economic activity for the future, as has been mentioned in section 2.1.3. But it has to be considered, that such development is in accordance with the protection aims of the sanctuary. At least 'protecting the nature' and 'being a touristic attraction' contain conflicts for the area, which have to be solved. In addition, the environmental awareness of local people is still weak. Therefore, from the perspective of conservation the three above-mentioned economic activities also can create threats in the future, namely
– Agriculture: intensification of fertilizers, extension of farmland
– Fishing: overfishing of the lake
– Gold mining: pollution of the soil and the lake, in particular with mercury, sedimentation of the rivers and lakes, location of some mines in the protected area.

9.2   HUMAN CAPITAL AND SOCIO-DEMOGRAPHIC SITUATION

Human capital

Following the endogenous development approach, qualified labour forces with specific skills and competences from the region instead of external and mobile ones are of an advantage and sometimes necessary (Tödtling 2011). Pike et al. (2006) also urge that in some areas weak or deficient education and skills among people and communities become the main barrier for successful development.

According to the results of the current situation of human capital, half of the population (50.5%) are working. Farming, fishing, government or private staff, casual work, other self-employment are the economic activities, which the people in working age (between 15 and 60 years old) are engaged in. As has been shown in section 4.2.2 in farming as well as in fishing the vast majority of the people working in this sector have only a low education with more than two third and three quarters have just monastic or primary school education. Only people employed in the government or in private companies do have a reasonable high education. Here more than four fifth of the employees have a higher education degree (at least high school).

The generally low education level in fishing and farming might have been sufficient for the traditional way of practising these activities, but considering the future development in technology and production modes the skills of the farmers and fishermen have to be improved, which only can happen by vocational training for the people already working in these sectors.

As has been reported in subchapter 4.1, almost a third of the population are students, which is a good potential for future development, if they get a high quality education (see subchapter 9.4) and if they can be kept in the area. In this context qualified job offerings within the region will be urgently required. This is missing and results in a brain-drain process.

Education

Concerning the current education situation of the area, some weaknesses can be highlighted. For instances, in some villages a problematic teacher to pupil/student ratio in particular in primary (1:46) and middle school (1:68) does exit. Moreover, some experts mention that in order to upgrade the quality of teaching capacity building trainings are urgently needed. Of course, such a sufficient and qualified manpower is necessary and is the basic pillar to produce the new qualified and skilful manpower/human resource, which is definitely vital to implement the endogenous development concept in the area. Simultaneously, the quality of higher education and diversified subjects should be offered in the region that could be one solution to control the brain drain process.

Nevertheless, according to experts currently one strength of the education issue can be mentioned. Parents in the area, especially young parents, are highly motivated to support their children's education. And they are also willing to contribute to some renovation of school infrastructure on a self-help basis organized by the village community. This shows their understanding about education, which is very positive and a good sign for the future development in the area.

Ethnicity

As has been already reported in subchapter 4.1, the ethnic groups, mainly Shan, Kachin and Burma, live harmoniously together without ethnic or religious conflicts. This unity is a good base for starting endogenous development. Moreover, since a few years ethnic language is allowed for teaching in school. This can improve the identification with the region, which increases the chances of more engagement of the people in endogenous development. But related to that one barrier has to be mentioned, too. Within the ethnic groups different dialects are spoken and in school it is difficult to decide which ethnic language and which dialect should be used. Besides that quite often the teachers are not able to speak the ethnic languages. Obviously, ethnicity can also be a cultural resource, it has a potential for tourism and ethnic niche economies, which can contribute to endogenous development.

Health

Concerning the current health situation of the area, the basic needs like manpower and health care facilities have to be covered in all villages (see subchapter 9.4). Besides that a big challenge can be mentioned in the area, namely the drug issue. Mostly young people between 18 and 30 years are drug users. It not only weakens the social and the health conditions of the population (family suffers) but it also has the consequence that these drug users are not able to work. Therefore, they cannot contribute to the economy and the society as well. In addition, the worst foreseeable consequence of drugs can be a serious health problem (e.g. HIV) in the near future.

In general, inhabitants do not have solid knowledge about health care. In particular many people still believe in traditional practices, and still depend on the treatment of untrained and unofficial medicine men.

Moreover, the hygiene awareness is low. Two fields can be highlighted: one is the behaviour in waste disposal (often the waste is just hauled outside the village) and the second is the general hygienic behaviour (latrine, drinking water), which especially in the rainy season results in many diarrhoea cases.

Migration

Currently a huge migration process takes place in the area, which leads to a kind of brain drain. On the one hand, young and well-educated people are leaving the area for education and work in other parts of Myanmar (e.g. in cities in central Myanmar like Yangon and Mandalay) and abroad (America, Australia etc.). Those people who stay in the area are not as highly qualified as the ones who move out so that local knowledge becomes weak.

On the other hand, unskilled labour and business-oriented people are immigrating into the area, who seek job opportunities in the gold and jade mining sector and other natural resources (e.g. fish). This reinforces a comparably high density of population, but it also has led to unstructured settlement patterns to some extent. The housings for the immigrants are partly located in the conservation area at the bank of the lake. Moreover, the resources are seen as belonging to the local population, which are now acquired and exploited by outsiders. It leads to some kind of alienation. The key question is who is profiting from the new developments. Very often, the benefit of local people is overlooked and they are decoupled from development, which causes tension and fraction.

The above described migration process weakens the human resource in the area in a double way. The substantial brain drain weakens local capacities, which could not be replaced by the in-migrants who have minor knowledge of the local society, assets and options. Neither do these immigrants have a local identity nor feel they responsible for social cohesion and regional development.

9.3   INSTITUTION OR ORGANIZATION

According to Pike et al. (2006) (see in section 2.1.1) traditional top down policies have been considered to have success in any specific case, and to have been transferred and implemented almost without changes in different national, regional and local contexts. Therefore, local institutional conditions were ignored to a great extent. Vazquez-Barquero (1999 and 2003) also points out that a balanced and integrated strategy can be achieved only by a systematic participation of local economic, social and political actors in the planning and development process. Their involvement is very valuable to achieve a careful analysis of the economic potential of any area. Thus, Assche and Hornidge (2015) urge that local and expert knowledge are essential for a good regional development policy. Moreover, decentralized decision making and policy competences at the local and regional levels is a favourable element for economic development, with respect to a better understanding of problems, barriers, and potentials for regional development as well as the needs and goals of the local and regional population (Tödtling 2011).

Due to the political system in Myanmar until yet the decision comes from above, that means that a top-down model is still being practiced. It is one of the barriers for endogenous development approach in Myanmar in general. However, since 2012 the government has started to consider creating more inclusive institutions of governing and to foster democratic elements at the state and regional level. Today, government also strongly focuses on the way of federalism, which every region and state in Myanmar needs. If this process will move forward, it can help that decisions can be done by the participation of all levels. Indeed, their efficient involvement can be only made on an adequate knowledge and experience base. Especially in Myanmar, these chances were lacking for a long time under regimes of socialism and military rule. In particular, in the Indawgyi Lake Area the quality of knowledge has been degraded because of the brain drain process, which has been reported already in subchapter 4.4. A way to overcome the just mentioned weaknesses have been addressed in the above subchapter 9.2.

Currently, in the area different institutions can be found namely governmental, anti-governmental and non-governmental institutions. As has been mentioned, the collaboration between governmental institutions often does not function properly. For instance, the department of fishery and the department of conservation do not work together hardly ever, even though it would be urgently necessary to reach the goals of conservation and protection. And an even bigger weakness and threat is the insecurity caused by the conflict between the anti-governmental institution, KIA, and the governmental institution, army. For instance, people and businessmen have to pay taxes to both sides. Moreover, this insecure situation is a main source of corruption and inefficiency. That means for example, the conservation team cannot patrol properly in the conservation area, and as a result some illegal extractions (gold, timber, firewood) take place in the conservation area. In addition, some development activities like infrastructure building or economic activity are blocked at least temporarily; in case of fighting the martial law comes into action.

For overcoming these mentioned problems peace between KIA and government army is urgently needed. The current government elected in 2015 restarted to organise the peace discussion with ethnic armed groups. Hopefully this time the peace process moves more efficiently forward due to the same interest, namely, the aim of KIA and of the new government is to establish a kind of federal system. So, in general both parties have the same ideas, but probably there are differences in detail. In the forthcoming talks and discussions it has to be figured out how the different positions can be brought together. Nevertheless, with respect to the endogenous development approach the common idea of federalism builds a good foundation.

Other challenges are manpower shortage and insufficient facilities for government institutions in particular in the field of conservation, education and health in the area. The situation for education and health will be described in subchapter in 9.4. Here the conservation topic will be in the centre of discussion. The conservation department is responsible to patrol the area, which is under protection. The patrolling for the lake seems to be fairly smooth due to the well organized fishing-free zone system and the support of some donors. In contrast, at present the patrolling for land looks rather complicated due to an uncategorized zoning system, insufficient facilities and unstable political situation. Moreover, the lack of a categorized zoning system causes conflicts between inhabitants and the conservation team. For instance, local people cut firewood and deposit waste. This is partly a traditional behaviour and shows minor environmental awareness. But partly this conflict is also generated by the lack of a clearly defined zoning system.

Probably, these problems can be solved considerably in the near future, because the zoning system is in process and the peace process is underway. Furthermore, recently the Indawgyi Lake has been listed as a Ramsar Site, consequently the governmental and inter-governmental institutions will take more intensive attention for zoning.

Of course, some positive issues also can be recognized in the area such as the strong involvement of some villagers in the controlling team and the establishment of community forest. Every village has a village community as has been explained in subchapter 7.1. This community actively participates in organising village life and village development. Such kind of knowledgeable and motivated local people are valuable specifically for helping to reach the conservation aim and in general for introducing and carrying out the endogenous development concept in the area.

Concerning the establishment of community forest one thing has to be mentioned: On the one hand it contributes to reforestation, fulfils the conservation goal, and supports a friendly environment. On the other hand, if the reforestation uses eucalyptus – as it often is –, it will harm the environment and eco-system, because this species consumes much water and is not endemic.

9.4　INFRASTRUCTURE

Vazquez-Barquero (1999 and 2003) points out, that insufficient infrastructures were seen as the root of the problems of many lagging areas by the previous regional development approaches. Therefore, most of the development policies were given the priority for the building of infrastructures. Also in endogenous development approaches infrastructure is a basic feature. But as Pike et al. (2006) mention the question is how and what kind of investments in infrastructure contributes to a sustainable development.

Of course, the research area has been an infrastructure bottleneck for a long time due to lack of security of the region and the periphery outback. This holds true for all sectors of infrastructures, like transportation, electricity supply, education and health facilities.

Transportation

Today the roads located within the area and the major transportation line between Mandalay and Myitkyina, which is bypassing the area, became upgraded by the government to be useable for the whole year (partly is finished and partly is under process). In 2015 a paved road in the west of the lake up to Nyaungbin and a gravel road in the east of the lake up to Lonsent can be found. This provides access to markets, to education opportunities and to a hospital in Mohnyin as well. Motor-boats are used on the lake as second transportation mode for inter-village connection especially in the rainy season. Mostly this waterway transport is used by the villages in the east of the lake due to the bad condition of the gravel road. In general the condition of the gravel road is not satisfying. Naturally, it is worst during the rainy season, but also during the dry season transportation is difficult because of the road holes.

As usual light transportation such as buses and cars are a private sector activity in Myanmar. Currently, in the research areas two private owned cars from Nyaungbin serve as regular transportation alternative. The frequency of the service (only once a day) is not sufficient and the transport charges are quite high. Therefore, many people have to take the taxi to have access within the area or the surrounding.

Although the situation is not always satisfying up to now the evaluation of the households for the future is optimistic when 82% see the future transportation situation positive compared to 40% at the present. And this evaluation is quite realistic keeping in mind the development in road construction in the last few years.

Electricity supply

In addition, public electricity supply for the whole research area is under process. Three villages namely Loneton, Mamomkai and Main Naung have already access to electricity. This access will definitely improve the living standard, and probably, it can also reduce the firewood consumption (currently all households use firewood for cooking and heating). Therefore, in the future the watershed area can be protected more efficiently and the conflicts between local people and conservation team will be less. In addition, it enables farmers to pump water from the lake and irrigate farmland on the fringe. It also opens employment opportunities: added value activities in particular in agriculture, fishery and mining like for instance, rice mill, cool storage, food processing and processing of gold and jewels.

Although this development in electricity supply is for sure a positive aspect, one also has to mention that in order to install the electricity line quite often trees have been cut, which is in contradiction to sustainable development ideas, in particular conservation goals. Nevertheless, future electricity supply is seen positively by about 75% of the respondents while only 15% of them evaluate the supply positively at current.

Water supply

In total 93.5% of the respondents depend on tube well/pump water for drinking, while the rest are lake water users. There are some critiques regarding drinking water quality especially in Lonsent and one medical doctor in Loneton also confirmed health risks. In the near future a high risk of water pollution caused by mercury contamination can occur. Indeed, access to clean water and sanitation reduces health problems and mortality. If there is electricity supply, it can decrease health risks by cooking the drinking water more easily. One more thing has to be mentioned here namely, that rainwater-harvesting practice is still lacking in the area.

Health facilities

Regarding health facilities currently only two 16-beds hospitals with one medical doctor each (one is in Chaungwa in the west of the lake, which is a bit difficult to reach due to the road condition and the other is in Loneton) are serving a population of about 50,000 in the whole area. In Loimon and Nanmon village respectively new hospitals like the one in Loneton are under construction. Moreover, health centres are located in Lonsent, Nammoukkam, Hepa, Hepu, Mamomkai, Main Naung, and Nyaungbin respectively with at least one midwife each.

Until yet no ambulance vehicle is available in the area. Transportation of patients has to be done by community owned cars, which naturally do not have proper first aid equipment. In particular, this lack is a big problem for patients, who are seriously ill and have to be transported to the Mohnyin 100-beds hospital. Another shortcoming related to transportation is that the medicine, which is free of charge, has to be transported from Mohnyin by a car, which has to be rented by the above mentioned health centres. Moreover, the cooling storage, in particular of vaccination, causes problems because of the lack of electricity supply. This problem might be solved in the near future due to the establishment of electricity supply.

Positively can be mentioned that the government is aware of the problems of health supply in the area. One obvious result is the enhancement of the manpower in Loneton hospital (one medical doctor plus two nurses in 2014 to one medical doctor plus five nurses in 2015). This is in accordance with the general increase of expenditures in the health sector in Myanmar (in the 2010–11 budget year the government health expenditure was 0.2% of GDP, two years later in 2014–15 it was almost five times more and resulted to 0.99% (Ministry of Health 2014)). Because of these new developments, the situation of future health facilities has been evaluated positively by twice as many respondents as it is the case for the current one (current 25% to future 50%).

Education facilities

Concerning current education facilities, it has to be mentioned that every village in the area has a primary school. Official middle schools can be found in Loneton, Nyaungbin and Main Naung and a high school is located only in Loneton. However, the local authorities (e.g. township officer, the head of the village), village community, teachers and officers from the relevant department organise to extend the school grade at the primary and middle school in accordance with manpower and infrastructure (the so-called school branch). As a result students do not need to relocate for visiting a middle school and similarly, a certain amount of student can attend the high school level without relocating. Undoubtedly, it is not an optimal solution, because the infrastructure shortage (classroom, latrine) and insufficient number of teachers (for example: a primary school teacher takes care of middle school level class and a middle school teacher works for high school level class) occur quite frequently. This causes a quality problem, which leads to the result that some students from the research area attend private schools in Mohnyin and Hopin for high school level education. This of course costs money, which cannot be afforded by many students. Additionally, a full high school education can only be received in Loneton, so students from other villages have to commute every day, which is extremely difficult (no public transport) and in the rainy season almost impossible. The result is that quite a number of students quit before going to high school. When the road construction will be further improved as it takes place in the last two years, the transportation problem will be reduced. But nevertheless it is urgently necessary to extend the number of high schools in a way to overcome the distance problem. One solution might be to upgrade the middle schools in Nyaungbin and Main Naung to full high schools and to consider upgrading a primary school to a high school at the east side of the lake. Then, four high schools would be existent, one at each side of the lake. Such a location structure has the advantage that all students from the villages around the lake would have reasonable distances to a high school.

According to the expert interviews, it is intended to upgrade the Mohnyin Degree College into a university. This for sure would be a good opportunity for students of the Indawgyi Lake Area, because it is nearby. And it also can be a chance to reduce the brain drain, which has been reported in subchapter 4.4. However, such a reduction can only be set into action successfully if two preconditions will be fulfilled: firstly, qualified job opportunities are offered in the lake area (e.g. in eco-tourism, conservation and research), secondly, regionally based study programmes are offered at the university. These new improvements only can be achieved by a general increase of expenditure in the education sector in Myanmar. Luckily, this is under way for some years (in 2011–12 budget year the government education expenditure was 0.8% of GDP (World Development Indicators 2013), one year later in 2012–13 it already went up to almost 1.5% of GDP (UNICEF Myanmar 2013)). This might be part of the reason that the situation of future education facilities has been positively evaluated by twice as many households as for the present (current 35% to future 70%).

Communication and information

The communication infrastructure based on mobile phone has been improved remarkably in the last couple of years in particular since the prices of the SIM cards have fallen (in October 2014). However, the infrastructure for information is comparably weak. For instance, there is almost no connection based on internet (which, in particular, will be a threat for the future development of businesses). Additionally, print media are also rare until yet because of the transportation bottleneck.

9.5   ECONOMIC STRUCTURE

According to the endogenous approach, economic growth of the area is taken from a long-term perspective and has to take into consideration the ecological and environmental situation of the region (Tödtling 2011). Obviously, economic development does not only mean the increasing of regional production and average income per capita, but also enhancing the broader socio-economic situation, in particular the living condition of the localities (Pike et al. 2006).

The current sluggish economic development of the Indawgyi Lake Area arises mainly from the peripheral location and the unstable political situation. Therefore, the area has difficulties to provide adequate job opportunities; however, it is obvious that job opportunities have to be generated on the base of the potentials (natural and human), the area has.

Agriculture

As has been reported in subchapter 6.1 some weaknesses of the current agriculture sector in the area can be mentioned. For instance, farmers in general practice the traditional way of farming, they have to deal with a lack of infrastructure for farming (farming machine and farm water availability) and of resources (money) and their knowledge about (new) agriculture techniques (especially high-quality species, multi-crop system and efficient cultivation techniques) are weak. Since 2012 the government provides short-term or seasonal loans of a low amount and agriculture specialists can share the knowledge about farming systems and techniques in some but not all parts of the area. But in future adequate and long-term loans and sufficient agriculture specialists to cover all areas are urgently required. This solution can attract farmers to keep their job and reduce the fishery and gold mining. Only the practicing of a multi-crop system with two harvests per year can make a better balance of the soil nutrients in the west of the lake (currently farmers in the west are facing a soil quality problem).

In the past, rice cultivation has depended only on the weather. Because of the climate change irrigation starts to become necessary at least for some parts of the fields. Here a problem arises which is caused by gold mining. This activity not only is responsible for the pollution of water with mercury, but it also leads to sedimentation and reduces the amount of water in the streams as well as the quality of the water (muddy water). This results in lower yields.

Moreover, according to the results the problem of too small farmland sizes can be highlighted especially in the west of the lake (half of farmers own less than nine acres). These circumstances result in remarkable lower income of the farming households west of the lake.

To improve the agriculture sector effectively the size problem is also necessary to take into account for future development. Additionally land use rights of the farmers are not always clarified (efficient allocation of the wild farmlands to farmers is necessary and important, which is still pending due to unstable political situation).

Currently, farmers cannot always find attractive and stable markets due to a lack of market accessibility. Furthermore, farmers can produce only low quality products because only a small-scale rice mill is available and very little food processing is carried out due to lack of electricity supply in the area and the accessibility problems. For instance, quite often paddy is sold to markets outside the area (Hopin, Mohnyin, Hpakant, Myitkyina) instead of rice. Moreover, vegetables are not cultivated very much because they cannot be processed further. Insofar, in the near future, when the road construction and the electricity supply are completed, these mentioned problems have a chance to be solved considerably.

Until yet, one current strength in agriculture is that the usage of fertilizer is comparatively low. This results in lower yields in particular in the west of the lake. But, such a low use of fertilizer also (a) is an indication of environmental friendly behaviour (Chambers and Conway 1991), (b) produces more healthy food and (c) can contribute to organic farming in the future.

Fishery

According to the results reported in subchapter 6.1, some current strengths of the fishery sector can be highlighted. It makes the second highest income for the area after agriculture. And the fish are appreciated by consumers and have a strong market even outside the area for instance in Myitkyina and Hpakant. That results of course in intensified fishing activities with sometimes using illegal fishing methods (e.g. electrofishing), very small mesh size fishing gears (e.g. lower than 0.75) and ignoring the closed season (from June to August).

Although the market demand for fish is high, the fishermen cannot make the highest-possible profit, because no further product processing is possible due to lack of electricity; only fresh fish can be sold.

Currently according to the conservation idea fishing-free zones are implemented and a rebreeding programme with domestic species has been started. These initiatives lead to a good future outlook, but some fishermen do not follow the zoning regulations. Insofar the good conservation plan is not as efficient as possible. Moreover, the water pollution caused by gold mining, in particular mercury and sedimentation, threatens the quality of fish.

A way to overcome these problems – for at least parts of them – can be environmental education programmes. As has been mentioned already for agriculture also for fishing a micro-finance programme is urgently needed.

Gold mining

No doubt, gold mining can create a good job opportunity for unskilled persons and a good income source for the mine owners at least the first two or three years after investing. But then quite often the mining is not efficient anymore. Insofar, gold mining builds a somewhat unsustainable income source. And very often the mine owners cannot go back into their former activity because they are in debt.

Moreover, this economic activity has severe negative social and environmental impacts. Almost all experts confirm that the drug problem in the area is caused mainly by the mining activity. And also the malaria disease is spreading specifically in the mining area. The environmental impacts are in particular: landslides happen quite often due to heavy rain and open pits; even more severe is the already mentioned mercury contamination, the blocking of streams and the sedimentation of the lake. This has adverse effects on agriculture and fishing.

Additionally, some mines are located in the conservation area. At the moment this cannot be prevented due to the unstable political situation. Within this context it has to be mentioned that natural resources like gold and timber and their extraction are reasons for the conflict between the two parties (KIA and army), because it is a good source for making profit and keeping or extending the power.

Potential economic activities

As has been already mentioned in sub-chapter 6.1, eco-tourism can be one of the promising future activities. Currently almost no tourism does exist, only two guest houses and one homestay are available. Due to the infrastructure situation and in particular the political one the area does not attract tourists until yet. Assuming these problems have been solved, some other challenges do exist to establish tourism. For instance, the standard of hygiene (e.g. quality of toilet, waste disposal) is comparatively low. According to the results, local people want to be involved in eco-tourism, but quite often they do not have an idea how to contribute. Additionally, the knowledge of local people has to be improved (e.g. dealing with tourists, acting as guides). Moreover, eco-tourism could be combined well with farming and fishery in a sustainable way by serving healthy local food (e.g. organic food).

However, one challenge concerning the relation between eco-tourism and conservation has to be pointed out clearly. The carrying capacity of the nature has to be respected, which leads to a limitation of the number of the tourists. The kind of mass tourism, like it has taken place at the Inlay Lake, has tremendous negative effects and would contradict the conservation goal and eventually destroy the environment. Only if this can be organized adequately future threats to the environment caused by tourism can be prevented and tourism can become a fourth economic base of the Indawgyi Lake Area.

9.6   SUMMARIZING – THE SWOT TABLE

The following table 9.1 gives a summary and an overview of the current weaknesses and strengths as well as the future opportunities and threats as they have pointed out and discussed in this chapter.

Table 9.1: Summary of the SWOT analysis

Topic	element	Present		Future	
		Strength	Weakness	Opportunity	Threat
Natural resource		• Good base for agriculture and fishing • Gold • Specificity of nature and beauty of landscape (including flora and fauna; esp. water birds)	• Lack of environmental awareness • Causes unstable political situation	• Eco-tourism	• Sub-optimal land use, fishing, mining and tourism management • Causes unstable political situation
Human capital and socio-demography	Human capital	• Young population	• Low qualification of people in basic economic activities (agriculture, fishing) • Brain drain	• Young population	• Not enough qualified job opportunities
	Education	• Interest of parents for education	• Insufficient supply with higher education • Insufficient manpower • Lack of vocational training for teachers • High number of high school student quit the school		
	Ethnicity	• Harmony	• Deficits in learning ethnic language in school	• Awareness, ethnic languages are important for identity	

Topic	element	Present		Future	
		Strength	Weakness	Opportunity	Threat
Human capital and socio-demography	Health		• Awareness of health • Low Hygienic standard • Drug problem		• Serious health problems (HIV)
	Migration		• Brain drain of qualified young people • In-migration of less qualified people • Causes tension and fraction		• Exploitation of natural resources • Brain drain • Causes tension and fraction
Institution/organization	Political dimension	• Involvement of village commu-nity	• Insecurity of conflict between KIA and army is blocking activi-ties (economic and protecting)	• Idea of federalism	• On-going fighting between KIA and army
	Admin-istrative dimen-sion/con-servation		• Decision comes from above • No collabora-tion between governmental institutions • Insufficient zoning system for conserva-tion on land • Lack of power full law, manpower and facilities	• Ramsar Site (internationally acknowledge)	• Suboptimal reforestation management
Infrastructure	Transpor-tation	• New paved road west of the lake	• Problematic gravel road east of the lake • Only few transport mode, low transport frequently and expensive	• Improving road conditions	

Topic	element	Present		Future	
		Strength	Weakness	Opportunity	Threat
Infrastructure	Electricity supply		• shortage of electricity supply	• Improving electricity supply • Enhancing living standard • Effective protection of watershed area • Support irrigation system and added value issue	
	Water supply		• Water quality not always satisfying • Lack of rain water harvesting practice		• Water quality pollution by mercury contamination
	Health facilities	• Community supports some facilities (car and building)	• Lack of hospitals • Lack of qualified manpower • Lack of cooling storage for medicine and transport, (e.g. ambulance)	• Improving the manpower and hospitals. • Improve some health facilities due to improvement of electricity supply	
	Education facilities	• Establishing of school branches for middle and high school	• Low number of high school • Insufficient building and qualified manpower	• Extending high school capacity • Upgrading Mohnyin Degree College to a university	
	Communication and information	• Good supply of people with cell phones	• Almost no internet connection		• Lack of connection for new businesses
Economic structure	Agriculture	• Good income source based on paddy • Low use of fertilizers	• Small farm sizes in particular west of the lake • Monoculture system	• Reorganizing farm sizes • Introducing multi crop system • Added value products • Better market accessibility	• Degradation of soil quality due to monoculture system

Topic	element	Present		Future	
		Strength	**Weakness**	**Opportunity**	**Threat**
Economic structure	Agriculture		• Insufficient knowledge of new techniques, infrastructure and investment • Lower yield due to muddy water • Lack of attractive and stable market • No value added process	• Healthy food/ organic farming • Financial assistance	
	Fishing	• High quality of fish • Good market for Indawgyi fish • Well designed zoning system	• Intensified fishing • Using illegal fishing methods • Ignoring closed season and zoning system • Lack of added value process	• Rebreeding program	• Overfishing • Quality of fish due to gold mining activity (e.g. mercury contamination)
	Mining	• Reasonable income source • Create job opportunities	• Negative impacts on environment (water pollution, mercury contamination, land slides and sedimentation) • Social problems (e.g. drugs, health and debt due to unstable income)	• Reasonable income source	• Negative impacts on conservation, agriculture, fishing, and human well being
	Eco-tourism	• Good potential, until yet not existent	• Lack of awareness and facilities for tourism	• Source for job opportunities (e.g. guides) • Income source for locals (e.g. accommodation, restaurants, food supply)	• Negative impacts for conservation goals if too many tourists

10   FUTURE DEVELOPMENT IN THE REGION

Possible development paths

An effective regional development does not work with only the endogenous forces. It is always the outcome of both endogenous and exogenous factors and processes and their interaction. In addition, regions normally possess a different capacity for endogenous development, and insofar they have different needs for help from the central government or for external assistance (Tödtling 2011). This leads to a diversity of development paths and development models.

In this context endogenous, but non-indigenous factors in particular infrastructure investments (especially in transportation facilities like road, communication and information facilities, electricity supply, and education facilities like schools, training organizations, universities and research organizations), are purposely created or upgraded by policy makers and related institutions. As a result, market accessibility, highly educated workforce, knowledge and technologies are developed in the region that might drive new products, processes or other new solutions. However, keeping in mind the critique of Pike et al. (2006) that heavy investment in infrastructure leads often to little or no emphasis on the other development factors such as the support of local firms (funding issue e.g. micro finance program/loan), of local human resources (vocational training, research fund) or technology transfers and spillovers. Therefore, a comprehensive and balanced strategy is necessary, which means: the improvement in infrastructures should be in accordance with the boost of local economies and enhancements in labour skills. Of course, similarly, inward investment also has to match the aim of improving the local economies, local labour supply and local infrastructure.

This chapter will not elaborate such a strategy completely but will only focus on some important elements and ideas, which might be part of a development concept. By doing so, the pros and cons of some development activities will be pointed out. The discussion will also consider that a comprehensive strategy has to keep in mind two sub-strategies of development, which are characterized by their temporal dimension:

1. Short-term perspective: the starting point is the current situations of the Indawgyi Lake Region, which needs immediate enhancements and help in different fields (e.g. vocational training for farmers, waste disposal management).

2. Long-term perspective: this perspective focuses on the preconditions, which have to be created in order to achieve a development, which is of long duration for the future. Establishing a sufficient higher education system in the area might be an example. This has not immediate effects for improvements of the economy and society, but it will lay a base for the education level of the future generation of people in working age.

As can be seen in the diagram of figure 10.1 both the perspectives differ in the time
dimension although activities can be and probably often have to be carried out
parallel. The difference is the time period for reaching the aim. In the figure this is
demonstrated by the two arrows. The broader (shorter) one represents measures
with short-term perspective. Measures have to reach their goals immediately or
after a short while. The trimmer and longer arrow represents measures with a long-
term perspective. The arrow becomes more intensively grey with on-going time. It
shows that the final aims of the measures are reached more and more.

Figure 10.1: Temporal dimensions of development

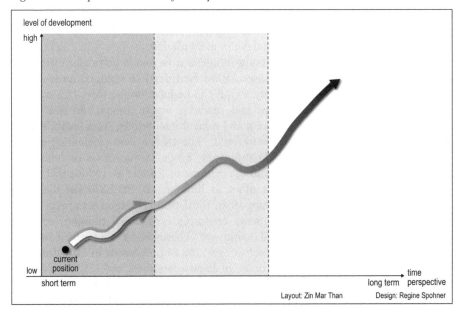

But whatever concept will be established and carried out the most indispensable
precondition is related to the political arena, namely that peace between the central
government and the ethnic armed group (KIA) has to be settled. Without peace and
a certain amount of autonomy at all levels, so-called federalism, all development
measures will stay weak or even will not work successfully. This political precondition has to be completed as soon as possible by the top leaders of both institutions
and is insofar more a central and external action, which has to be taken before an
endogenous development concept for the area can be implemented with the expec-
tation of success.

According to the outcomes of research and discussion reported in chapter 9 the
following socio-economic potentials can be mentioned in the area in accordance
with the concepts of endogenous and sustainable development theory:
1. Human capital/resource,
2. Agriculture,

3. Fishery,
4. Mining,
5. Eco-tourism.

Human capital/resources

Generally, as has been reported in section 4.2.1 and subchapter 9.2, the majority of the people in the working age attained only the monastic or primary school level education. Only about a third of the people attended middle school education. Quite few people (less than a fifth) had high school or university graduation respectively. The main reasons for the low education level are that in the past only the primary school was available in every village and that a higher education for young people was blocked by transport bottleneck.

Generally, the low education level is a satisfying base for the traditional way of doing respective activities (e.g. fishing, farming and mining). However, considering the further development in each activity vocational trainings for the people already working in the sector are necessary to improve the situation immediately.

Beyond this some training on matters like leadership skill, community participation and team building are also important aspects, which have to be implemented soon, too. This is in particular necessary for carrying out an endogenous development approach, in which some forces like social and political factors in particular involvement of social agents and civil society, which stimulate processes of self-help, local initiatives, and social movements intending for the improvement of living conditions in a particular region play a fundamental role. In this way local forces and factors are strongly involved in a development strategy and it is intended for the needs and objectives of the regional population (Tödtling 2011).

Today, the situation of the school infrastructure has improved already. Nevertheless, one has to keep in mind that one third of the population is under 15 years old, which in particular means this people will be the coming local human resource for the endogenous development. This human resource should have qualified education, which obviously leads to the consequence that the school facilities (teacher, building, teaching media, upgrading of school) have to be improved further on. Some suggestions for the local basic education structure have been given already in subchapter 9.4.

Also good facilities are necessary for higher education of the local people. Myitkyina University and in particular Mohnyin Degree College, which are not far away, are good opportunities. However, one has to think about new programmes, which better fit to the locals' needs and are aiming to give a region-based knowledge to the students.

For instance, Cologne University (Germany) offers programs to the students to get in-depth knowledge on regional based study (e.g. Cologne, the Rhine area, the mountains, the flora and fauna of the region). Thus, the students have a deep regional based knowledge. In general, this chance is lacking in Myanmar until yet because knowledge is not made ready on a regional base. Therefore, this concept of Cologne University could be taken as a model to improve local skills in the region. In this context Myitkyina University and in particular Mohnyin Degree College would play a key role. Mohnyin Degree College can offer students regional based study such as zoology or botany of the region/Indawgyi area, geology of area, etc.

Indeed, universities can help giving students options and knowledge and this is part of the responsibilities universities have. With respect to future development such an education has a number of multiplier effects, which support an endogenous development strategy. In particular, the following ones have to be named:

1. Education is closely interrelated with economic development. In general one can say, the higher the education standard, the better the whole situation is becoming. And the more the whole economy is stable, the more options are there namely good job opportunities.

2. Because of this result and the general wish of people to have a good and secured life, a decreasing number of people want to be involved in conflicts. Thus, a good education can be a part of conflict solution.

3. Besides the above mentioned economic relation to conflict solution another dimension have to be mentioned. A high quality education makes the people ready for looking to issues from different perspectives and for sharing knowledge with each other. Therefore they can participate in discussing questions like 'What is Kachin?', 'What is good?', 'How to develop?', and 'Which are the priorities?', which are fundamental for implementing and carrying out a locally based endogenous development approach.

4. Moreover, the role of university is to help giving the knowledge to the local people to stay in the area, so they can have an own development. The new-comers need to see that they have to accept that there are local people and they have to arrange with them, not just exploit them. Similarly, in Germany the regional development policy is balanced and strengthens the local people with their area knowledge. Therefore, Mohnyin Degree College can change the situation by developing the educational mind set for the new generation and keeping the education people for the region, and encouraging the departmental research in the region.

Agriculture

According to the regional potential improving the agriculture sector is one of the basic elements for a successful endogenous development approach. As has been already mentioned in subchapter 6.1 and 9.5, the current weaknesses of the agriculture sector of the area in particular are the following: infrastructure resources (e.g. farming water availability, market accessibility, machinery, money), knowledge about farming systems and techniques, the natural condition (esp. the nutrient of soil) and the structure of the farming (e.g. small farm size).

Undoubtedly, infrastructure resources and the structure of farming can be improved much quicker than the others factors like knowledge sharing and improving the natural condition. Such a kind of education process will take time (so called long-term process) and is more difficult to carry out than other infrastructure improving issues.

For instance, infrastructure resources such as farming water availability can be improved very soon by doing a rainwater harvesting programme or implementing the electricity supply, which is underway. Similarly, market accessibility also can be enhanced through better transportation facilities, as has been already mentioned in subchapter 9.4 and 9.5. Moreover, the current government intensively takes into consideration to solve other issues like lack of machinery and short term and low amount of credit loans. These improvements of how much loan and how long it can be borrowed are under process. In this case, it is very much necessary that the financial supports are efficiently used by the farmers according to the purpose of the programme. Parallel the government has to make sure that enough money is left for the expenditures of vocational training and does not only invest in infrastructure issues.

Considering the vocational training related to farming systems and techniques one thing has to be kept in mind. It has to be considered that the educational level of the current farmers is only low and therefore proper intensive training courses have to be implemented. In this context practical examples are probably a good incentive to show the farmers how it can be done (fields with new rice species of high quality, new efficient cultivation techniques, applying new machinery and proper use of fertilizer and so on). Obviously, such courses have to be repeated frequently and should be held all over the area, so that knowledge and techniques are deeply rooted in the farmers' mind. Of an advantage also would be if a monitoring and advisory system can be established.

As already mentioned in subchapter 6.1 and 9.5 using the multi crop system can help to increase the income of the farmers, in particular at the west side of the lake. Additionally, such development improves also the nature condition in particular balancing the soil quality and it prevents the overfishing issue, because the farmers can work all year long in the farming business. Until yet second crops have been grown mainly east of the lake in particular peanut and soy bean. This of course, should be extended into the direction of vegetables (e.g. cucumber, maize) and fruits (e.g. pineapple, melon). But a necessary precondition is that a sufficient demand for these new crops does exist. One opportunity is food processing and exporting the products to urban market. Another opportunity could be to use the crops and products in the newly created eco-tourism sector.

Concerning the structure of the farming one weakness, which has to be solved in the future, is the farm size, which is too small in particular west of the lake. In general, three ways of extending the farm size can be mentioned:
1. Establishing new farmland in the area: Such new farmland can only be established in areas, which belong to the conservation zone. Insofar, this opportunity is not possible at all because it contradicts the conservation goal.

2. Using wild land to extend the existing farm: In the area exists a certain amount of wild land as has been mentioned in subchapter 6.1. According to experts some of these farmlands are already reserved for villagers, who do not own farmland and want to become a farmer. So wild land might be part of the solution for extending the farm size, but this land resource probably will not be sufficient.
3. Reducing the number or farmers and combining the fields: In the long run a high probability exists that several farmers will give up farming in particular if they can find alternative income opportunities. The eco-tourism sector might offer such kind of opportunities in the future. The land of these former farmers can then be used to extend the farm sizes. To do this for each village in an efficient way a farmland consolidation program is necessary. In such a programme all farmers of the village and the policy makers (e.g. agriculture and conservation department, village community) should participate and collaborate intensively together. This will be a long, and not always easy and successful process. Nevertheless, it is an integral part of endogenous development.

Fishery

According to endogenous development a strategy based on natural resources in particular fish can be one of the sectors for economic development in the area. As has been discussed in subchapter 6.1 and 9.5, this sector supplies traditionally the second highest income for the area besides agriculture. Considering the further sustainable development of the fishery sector the current weaknesses should intensively take into account. In particular the behaviour of the fishermen with using illegal methods and fishing gears, and ignoring the closed season and zoning regulation (fishing-free zones) must be improved through awareness and education programmes.

As has been already mentioned for agriculture, one thing has to be kept in mind for the fishing sector. Almost all fishermen only have a low educational level and therefore proper intensive education programs have to be implemented and these programmes have to be repeated frequently and should be held all over the area, so that knowledge is deeply rooted in the fishermen's mind. Of an advantage also would be if a monitoring system could be carried out continuously in future. Moreover, indeed, it would be very good, if either micro finance program or creating alternative income opportunity for fishermen's households could be established. This will result in overcoming the intensified fishing in the lake. One alternative income opportunity for fishermen can be a further processing of the fish and exporting these products to urban markets. A second opportunity is to take a job opportunity in the newly created eco-tourism sector (e.g. boat rental or guide for boat trips). In case of eco-tourism development fishermen can find an extended market for their products (ready-made or raw material) in the region itself.

Mining

This activity is based on the natural resource gold, which is quite unique for the area. Insofar, it can be one of the potentials of the area for implementing an endogenous development concept. Without doubt, this activity causes a lot of problems (e.g. pollution, landslides, drugs abuse) for the environment and society. Oppositely to these weaknesses the mining sector also has a strength insofar that it offers job opportunities, which means it is one of the income generators for local people. From the pure conservation or environmental perspective it would be clear to shut down this economic activity, but that would mean to eliminate one of the few basic sectors, which the area has. At the moment such a loss seems to be too severe when introducing an endogenous development concept. The consequence is that one has to think of more environmentally sound extraction processes, which for instance means using mercury in a proper way, and controlling the emission of mercury into the environment. According to the environmental law the mine owners have to be more carefully about the waste disposal and the refilling of vacant open pits. A third task is that at least within several years the mines, which are located in the conservation area, have to be closed. All these measures induce additional costs, which probably a small-scale mining enterprise cannot bear alone. And insofar, public financial assistances (loan) should be given to the enterprises. This can lead to a compromise between an environmentally risky activity and the conservation goal of the wildlife sanctuary. Moreover, in combination with eco-tourism mining can be an attractive tourist destination, if the above-mentioned improvement measures have been done.

Eco-tourism

As has been already mentioned in subchapter 6.1, the unique nature of the area (lake, landscape, migratory birds, fish, ethnic tradition, availability of hiking, etc.) is a potential and might be an attraction for tourism. Indeed, this economic activity will contribute to the creation of new job opportunities outside of agriculture, fishery and gold mining. Of course, it can also strengthen the activities in food processing and compensate the often declining agriculture and fishery economic activity due to climate change and closed season of fishing (Telfer 2002: 147). Even when this potential of eco-tourism is acknowledged as a factor for the area development it will be difficult to concretely measure its economic impact in detail, because tourism is dependent on a multi-dimensional factor set. In general the success of tourism is strongly depending on regional conditions (i.e. not only the area uniqueness, but also area security and so on) and on requirements for touristic attractiveness such as accommodation, food and beverage, service (including information), infrastructure, character of the village, natural prerequisites, traffic and accessibility (Sustainable tourism for development guidebook 2013: 194, Neumeier and Pollermann 2014). Obviously, it is not sufficient only to consider the local attractions, rather it is also important taking into account the tourists' demands (Neumeier and Pollermann 2014: 275).

Nevertheless, tourism does not only induce positive effects but might also have negative impacts. For instance, tourism development can negatively impact the eco-system (by ignoring the carrying capacity of nature, see subchapter 9.5) and some-times it can also contribute to an increase in conflicts among the local population, tourists and tourism operators (for instance when overlooking the interest of local people). Therefore, a suitable way to deal with the constraints and opportunities in practice is crucial. When establishing eco-tourism the following four aspects have to be considered in particular (cf. figure 10.2):

1. offering nature,
2. locally based economy,
3. locally based attitude,
4. protecting nature.

Figure 10.2: Development of eco-tourism in the Indawgyi Lake Area

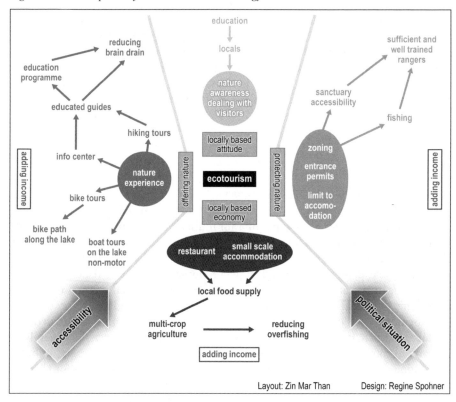

Offering nature

As has been mentioned before the nature of the area will be the most important element for establishing eco-tourism. In order to give the tourists a good experience of the lake area and its nature at least four potential services should be mentioned here:

1. Info centre: an info centre is a good measure for giving a good overview and introduction into the nature (situation, problems, etc.) of the area and also into the socio-economic situation. Such a centre should also offer information material (brochure, map, accommodation list, rental opportunities, food locations, etc.) as well as guided tours and presentations.

2. Boat tour: the lake is undoubtedly a very central touristic element and boat tours on the lake are a unique attraction for visitors, like it is also the case at the Inlay Lake. But in contrast to the situation at the Inlay Lake it is suggested that not motorboats are used but only boats driven by hand (rowing, kayaking) or electric power (battery, solar energy). In this way boat tours are much more environmentally friendly (reducing noise, water and air pollution, reduced waves) and suitable with the conservation issues.

3. Hiking: hiking tours can give good opportunities for tourists to get detailed outside experiences of the nature (flora, fauna, forest, agriculture land, etc.). Guided tours should be offered, but also self-guided tour trails, which do have outlook points for the scenery and for bird watching and signs with information on the location (e.g. geology, flora and fauna, history), can be a good opportunity.

4. Bike tour: in particular the shore area of the lake offers a good opportunity for bike tours. An optimal solution is an at least partly separate offroad path (for biking and for hiking). By doing so the tourists can enjoy the activities like hiking, biking or bird watching more intensely. Such a path should be quite near to the shoreline of the lake. Optimally the biking and hiking path circles the lake completely (until yet there is no path at the northern end of the lake). Here problems can arise related to conservation issues.

The tour guides and the personnel in the info centre should be well educated with deep knowledge on the locality (nature as well as society). Therefore, trainings and education programmes should be offered in the area or at Mohnyin Degree College. By this, good qualified job opportunities, which can have an influence on reducing the brain drain and which is a good resource for additional income, will be offered.

Locally based economy

Obviously, such kind of tourism will support the local economy (MCRB and HSF 2015). To strengthen the benefit option for locals as has been already described in subchapter 6.1, a community-based tourism or a local entrepreneurship with a small-scale enterprise approach can be established. If doing this in a cooperative manner, a number of people can participate and can have profit without having to invest much. Rental of cars, boats, or bikes as well as selling souvenirs can be income sources besides restaurants and accommodations. According to the idea of endogenous development small-scale accommodation run by locals should be established, because then it will be assured that the profit stays in the area and it fits best to a sustainable development. The restaurants as well as accommodation can be connected with local food supply (rice, vegetable fruit, fish etc.). In such a way multi-crop agriculture and fishery is supported. Because of the better situation of the agriculture, a reduction of overfishing might be an additional positive result, as farmers do not need an additional job. On the other hand, the threat of overfishing can arise due to an increase of fish consumption.

Locally based attitude

The tourism in the lake area is almost only based on the natural attractiveness. This has the consequence that the local people have to behave in a way that this natural attractiveness will be sustained. This means the people have to be aware of the nature of the area and have to act in a proper way. Here additional environmental awareness training is urgently necessary. Moreover, people have to be trained how to deal with tourists properly.

Protecting Nature

The aspect of protecting nature is vital and has to take into detailed consideration because the intact nature is the very base for tourism and the area is conserved as wildlife sanctuary. The protecting issue consists of two territorial categories, namely the lake and the land. Their carrying capacities have to be respected. Therefore, zoning systems are crucial for both categories. The zoning has to define clearly which activities in which intensity are allowed in the different zones. Here the land use 'tourism' has to be considered as a special one. For the lake such a zoning system has been established already (see subchapter 7.2). For the land such a system is in process, but not installed yet. The zoning systems are also necessary without introducing eco-tourism. Of course, if eco-tourism is established, such a zoning system becomes even more important. It needs categories of different access intensity for tourist. For organizing and running such zoning systems properly a sufficient number of well-trained rangers is necessary. This offers also new jobs for the area. Therefore new sources of income are available.

Moreover, when eco-tourism is established, it has to be considered carefully how intensely eco-tourism is allowed. For sure the area cannot bear a very high number of tourists because of the conservation issue. Therefore a limitation has to be set and controlled. In this case for instance, two categories of visitors have to be considered: day tourists and overnight tourists. The number of tourists in both categories can be controlled by collecting an entrance fee at the gates. According to the present street system at least two gates are necessary: one at the southeastern fringe of Nanmun and one at the northern fringe of Naungbin. This collected money can be used for conservation issues. Additionally, the number of accommodation facilities should be limited so that only a certain amount of overnight tourists are allowed. How high the number of permits per day is should be discussed and decided by the different levels of stakeholders. These stakeholders also should think of whether entrance fees are only taken for a person or also for a vehicle brought by the tourists.

At present, the limitation is not urgently needed because the number of tourists is very low. In so far the authorities can gain experiences with the growing number of visitors and based on these experiences proper limits for the future can be set.

Developing the economic sector 'eco-tourism' in the above mention way can only be successful if a stable political situation is assured and if the accessibility of area is improved. The latter has started (paved road in the west and gravel road in the east of the lake, paved road from Mohnyin and from Hopin to the lake area). But further improvement and maintenance are necessary.

11   CONCLUSION AND FINAL REMARKS

The research area is located in Kachin state, in the northern part of Myanmar. The region of 775 km, including an open water area of 120 km², is conserved as 'Indaw-gyi Wetland Bird Sanctuary' since 1999. It also belongs to the ASEAN Heritage sites and in February 2016 it was listed as Ramsar Site. In the area 95 species of water birds and in the lake 64 fish species are recorded, some of which are endemic and can only be found in this area. The climate is Savanna climate and the soil quality is in general suitable for agriculture. Due to these favourable conditions, a population of more than 50,000 are living around the lake and the area is quite densely populated. The people's livelihood entirely depends on the natural resources. The activities are farming, fishing and gold mining.

The socio-economic potential of the Indawgyi Lake Area was until yet widely undiscovered. However, the pressure of continuous population growth, growing market demand, the globalization trend, and lack of knowledge and infrastructure lead to unsustainable use of natural resources and endanger its long-term development opportunities.

The research analyses the present socio-economic situation of the Indawgyi area in depth and detail, as it is not only important for the local population, but has significant impact on the regional and national level. The aim of the research is to outline the potentials for socio-economic development of the region and its position in the broader context, to describe current threats to sustainability and environmental degradation, as well as to identify and describe possible long-term development outlooks that are environmentally and socially feasible.

11.1   REMARKS ON THE CONCEPT OF THE RESEARCH

Intellectually, this research is mainly based on the two concepts of endogenous development and sustainable development. The reasons to apply the concepts of endogenous and sustainable development are the followings:
– to implement bottom-up local and regional development policies for a so called tailor-made development approach,
– to use its regional uniqueness as a source of competitive advantages,
– to respect the local economic, social, political and institutional conditions,
– to apply local knowledge (i.e. in-depth understanding of the regional setting and situation can contribute to enhance the potentials of socio-economic development),
– to sustain the social and economic improvement of the area or region through the involvement of the local people,
– to respect the carrying capacity of the ecosystem.

In order to carry out such an endogenous development approach several factors are highly important as for instance: the improvement of the competitiveness of local firms (entrepreneurship skills, own branded products), the enhancement of human capital or labour skills (innovation skills), the building of infrastructure including modern information and communication, the attraction of inward investment and regional or area institutions.

This research agrees with the statement of Tödtling (2011), that an endogenous approach is more broadly defined in comparison with an indigenous approach. For instance, a pure indigenous approach hardly takes into account factors like infrastructure issues (road construction, building of school, university, vocational training, hospital, establishment of electricity supply and so on) and resources for vocational and education trainings, which come from the outside. So, a successful endogenous concept has to keep in mind that an effective or successful regional development is based on the outcome of both endogenous and exogenous factors and processes and their interaction.

Moreover, the research keeps in mind that the concept of endogenous development for a region is not an 'island of development'. Therefore, the development concept of this research is interrelated to a considerable extent with national political and institutional structures and with the global context as well.

In accordance with these above concepts and the necessities of endogenous and sustainable development approaches the key research questions were formulated. These are mainly grouped into three categories. Category 1 is related to the current situation (e.g. demographic, social, infrastructure, economic, governance and conservation status). Category 2 is dealing with evaluations (e.g. current and future socio-economic situation at household level and village level) done by the locals. Category 3 is related to possible development paths for the area; in this context ecotourism is highlighted and the solutions for present and future threats are also outlined.

11.2 REMARKS ON THE EMPIRICAL WORK

The empirical work is mainly based on two methodological approaches. The quantitative one uses a household questionnaire. In total 216 households in 10 villages were surveyed. The qualitative approach is based on interviews of altogether 54 experts of different fields such as farming, education, health, and conservation. Additionally field and participant observation have been carried out. Secondary data were used if available. But there are only very limited sources. The data collection was organized in two phases, which has the advantage that in the second phase extensions as well as corrections can be considered well to complete the necessary information. Both the approaches (quantitative and qualitative) have been combined by triangulation.

According to the findings and discussions, the recent economic and social developments in the Indawgyi Lake Area show several remarkable dynamics. In respect to potential regional improvements, the following recommendations can be made.

Concerning infrastructural improvements, an upgrading of the existing circular road connecting all villages surrounding the lake would help to balance out the currently different development levels between the east and west of the lake. The continuation of the impressive upgrading of roads has to be judged positively. As a consequence the road can be used for the whole year and the frequency of transport modes will increase, which creates a better accessibility to the area. It can result in the improvement of economic, health and education situation of the area.

Concerned with improvements of communication and information facilities better providers for internet connection would support the future development of the local economy. As a consequence people also have better opportunities to learn about and use modern technologies (techniques) and it also can contribute to build and improve the social networks.

The establishment of an electricity supply system and a sufficient and stable supply of power would lead to better living conditions, improvement of health care situation and a reduction of firewood consumption. It would provide a solution for the deficiency of farm water availability and the problem of water quality, it would allow a better communication and information as well as it would enhance the economic situation (for instance considering added value processing in particular for agriculture, fishery and mining products). Regarding water supply the practice of the rain water harvesting would contribute to solve the problem of insufficient farming water.

Regarding the health infrastructure, the present condition of ambulance, medical transport and cool storage is not satisfying. To improve this situation two ambulance stations, which offer the above mentioned activities, should be established in a first step: one at the east side and one at the west side of the lake. An upgrading of current existing health care centres and new establishments of 16-beds hospitals in Nanmon and Lewmon would help to cover the need of the whole area. In addition, the upgrading plan of the 100-beds hospital into a 200-beds facility in Mohnyin would enhance the opportunity for appropriate health care and physicians within a closer reach.

Moreover, adequate manpower for the health care centres and the 16-beds hospitals in the area should be taken into consideration seriously. Furthermore, health education programs for locals in particular with regard to drug issues, nutrition, hygiene and sanitation are not only necessary, but upgrading vocational trainings for personnel as well. The intensity of health education programs should be sufficient to overcome the traditional beliefs, which still do exist. Vocational trainings for personnel should be held in the region rather than in places outside the area. By doing so the degree of participation can be enhanced. To improve the health situation of the area not only the personnel from the health department do have responsibilities but also the involvement of all stakeholders (in particular, different level of authorities, local community and locals) is necessary.

With respect to the improvement of the education infrastructure the current existing primary schools especially in Lonsent, Nammoukkam, Mamomkai and

Nammilaung should be upgraded to middle school level and similarly middle schools should be enhanced to high school level in particular in Nyaungbin, Main Naung, and Hepu. In addition, some other facilities such as in sports, library and clean drinking water supply for students, which are still lacking, should be taken into consideration for improvement. At the same time sufficient manpower and the vocational trainings for teaching personnel are urgently needed to consider. Vocational trainings should take place in the region as mentioned above. Moreover, the well-being of personnel from education department should not be ignored, because this can really stimulate their performance (motivation, creation and innovation).

Furthermore, the school curriculum should be upgraded by including environmental education courses, which are still lacking, as well as ethnic language courses, which should be established in the first step, even though there are some obstacles inherent. Regarding higher education, an upgrading of Mohnyin Degree College into a University, which is still pending, and installing a research unit related to the regional based study program at the university (which does not exist yet in Myanmar) would help to get qualified and knowledgeable people for the region, who can contribute to develop the endogenous development concept for the region. To improve the education situation of the area all stakeholders in particular parents, students, personnel from the education department, and all levels of education institutions need to participate.

Concerning the imbalances in the migration patterns and in order to prevent the brain drain issue, the upgrading of schools and an increase of the number of high schools in the area are necessary. In parallel sufficient and qualified personnel for the schools are vital. Moreover, the above mentioned upgrading of Mohnyin University would help to keep the educated people in the area. In addition, job opportunities for qualified people in the eco-tourism sector can be foreseeable. Likewise, the improvement of quality and quantity of health care, education and higher education sectors can generate job opportunity for educated persons.

Looking at the ever-growing number of seasonal in-migrating fishermen and gold mining labourers only a good conservation regulations could control this problem efficiently as Myanmar has no specific internal migration law. That means a better control of fishing, in particular the ban of illegal methods, fishing-free zones and a closed season, is necessary. The comprehensive control of mining activities according to the environmental law and humanitarian standards is essential. In addition, in order to compensate the resources exploitation process cooperate social responsibility (CSR) programmes should be developed for the area.

Functional government services and institutions are one important pillar of the endogenous development approach. Currently, the most important issue is that a lasting peace has to be built up between the anti-government (KIO/KIA) and the central government, otherwise all development issues cannot be handled efficiently. It is positive that both parties intend to have a federal system, but the negotiations about a detailed structure of the system have only just started. It is foreseeable that these discussions will not be easy and will take time. For sure, decentralization and democratization of power will be part of the concept and have to be put into practice, which is the core concept of an endogenous development approach.

Consequently, every region can have its own development concept according to the region's context. For this it is necessary to have a good and balanced system of governance in order to formulate a sustainable regional development plan, based on multi-stakeholder decision-making processes in collaboration with the local leaders, who can carry out the plan in the long run. Therefore the strengthening of the already existing village communities and the availability and ability of all level of institutions are appropriate. Furthermore, current existing institutions in the area (e.g. conservation, fishery, education, health and so no) should be enhanced in quantity and quality in order to fulfil their tasks on the regional level and below. At the same time these institutions have to collaborate in a transparent and responsible way, which at present is quite often not the case.

Proper management of the natural environment of the Indawgyi Lake Area is probably the most important task for future development in order to preserve an intact nature and to keep its title as a natural heritage site. This is also important because future economic developments (e.g. eco-tourism) will rely very strongly on that. Hence the conservation issue becomes of high priority. Hitherto, properly organizing the conservation issues is facing serious challenges such as the unstable political situation, the un-clarified zoning system, insufficient facilities and manpower and lack of knowledge about waste disposal/management. The unstable political situation can only be solved by a peace agreement between the two parties (KIO/KIA and central government). To overcome the other above mentioned problems the involvement of all stakeholders (international, national, regional and local level) are fundamental in terms of strategies, technical means and financial resources. Concerning waste disposal management (in particular for mining and household waste) a powerful law and environmental awareness trainings are strongly recommended. Reforestation and fish rebreeding programmes should be enhanced in a proper way (species have to fit with the area).

The above mentioned infrastructure improvement, in particular transportation and electricity supply, would enhance the economic development of the area. The agricultural productivity and marketing potentials could further be upgraded to higher yields in addition to encouraging small-scale agro-industry to preserve food for enhanced export opportunities to other regions in Myanmar, such as the Mandalay area. Parallel, comprehensive education programmes for farmers related to multi culture cropping systems, quality species and knowledge of cultivation techniques should be conducted (keeping in mind that according to the household interviews more than two thirds of the farmers have only monastic or primary school education). Adequate financial assistance and availability of infrastructure for farming should be intensively developed and land consolidation programmes for agricultural land, which is divided into small plots should be taken into consideration.

Stronger regulation of fishing activities is necessary in order to preserve the unique fish stock. As a consequence, measures should be taken to reduce or control seasonal in-migration from outside the region. Proper education programmes for fishermen, micro-finance programmes and alternative job creation (keeping in mind that more than three-quarters of fishermen attained only monastic or primary school education) are essential. As already mentioned above the infrastructure improvement could enhance the potential of added-value processes, which would contribute to prevent overfishing.

At present, comprehensive regulations for mining procedures in the area are urgently needed to mitigate negative impacts on the social and environmental conditions as well as on the agricultural and fishery sector. As this economic activity offers job opportunities for unskilled people, sustainable mining procedures and added value products processing should be considered as a first step. Later this sustainable mining process could be combined with the development of eco-tourism in the area (e.g. mining tours).

Strengthening non-farm activities, such as the production of souvenirs, seems possible. A tourism base on sustaining the natural and cultural environment seems feasible provided that the use of the lake is strictly regulated and does not to exceed the carrying capacity. It is important to ensure the careful development of a community-based tourism, which will benefit the local population, rather than any kind of large-scale development that does not benefit the locals.

11.3 CRITICAL REMARKS AND SUGGESTION FOR FURTHER RESEARCH

Critical remarks

The quantity of surveyed households is on the margin for some of the statistical analyses which have been carried out. The set of questions and the structure of the questionnaire were adequate for the research topic and worked rather well. The questions covered all research topics, which have to be investigated. Nevertheless, one shortcoming was noted: the question 'how many acres does a farmer own' was asked, but an extended question like 'is the farmland connected or is it separated in plots in different places' has not been asked. But this question is an important one for investigating the situation of farming in more detail. Additionally, as usual in social research, some respondents hesitated to answer questions on their household incomes.

The number of expert interviews and the quality of field and participant observation were very good. The interviews of the experts of all fields were relevant for the research. In some interviews not one but more than one interviewee (from the same field) was apparent. This might have the advantage, that different (personal) perspectives and experiences can be recorded and enrich the information. However, according to the experiences of the author such a constellation was not that efficient, because the participants sometimes paid too much attention to each other and did not talk freely. Sometimes they also started to answer parallel.

Suggestion for further research

The study aimed at analysing and identifying the socio-economic potentials of the Indawgyi Lake Area. This has been done quite productively and can be considered as a first step to lay down a knowledge base for future development activities. This base should be enhanced by more detailed surveys of the agricultural land structure, a detailed analysis of quality of water and a detailed study for a concept of eco-tourism development. Additionally a monitoring system for the development processes should be implemented. This can help to react quickly and properly in cases of problems and changes, which might arise.

REFERENCES

Abdrabo, M. A. and Hassaan, M. A. (2003): A manual for socioeconomic study. From river catchments area to the sea: comparative and integrated approach to the ecology of Mediterranean coastal zones for sustainable management (MEDCORE). Accessible at: http://www.medcore. unifi.it/socioeconomy_manual_cedare.pdf (15.10.2013).

Alvesson, M. and Kärreman, D. (2011): Qualitative research and theory development. London: SAGE Publications.

Amin, S. (1976): Unequal development: an essay on the social formations of peripheral capitalism. Sussex, England: John Spiers.

Arino, O., Perez, R., Julio, J.; Kalogirou, V., Bontemps, S., Defourny, P. and Van Bogaert, E. (2012): Global Land Cover Map for 2009 (GlobCover 2009). © *European Space Agency (ESA) & Université catholique de Louvain (UCL)*.

Assche, K. V. and Hornidge, A.-K. (2015): Rural development: knowledge & expertise in governance. Wageningen: Wageningen Academic Publications.

Bärwaldt, K. (2016): Mehr Demokratie in Myanmar wagen. Herausforderungen für die Regierung unter Aung San Suu Kyi. Accessible at: http://library.fes.de/pdf-files/iez/12558.pdf (12.08.2016).

Bhat, A. R. (2005): Human resource and socio-economic development in Kashmir Valley: a geographical interpretation. New Delhi: Dilpreet Publ. House.

Bernard, H. R. (2006): Research Methods in Anthropology: qualitative and quantitative approaches. Lanham, Md: Altamira Press. 4. ed.

Bernard, H. R. (2013): Social research methods: qualitative and quantitative approaches. Los Angeles: SAGE Publications. 2. ed.

Bhandari, B. B., Nakamura, H. and Suzuki, S. (2015): Indawgyi Lake: the one and only tectonic lake in Myanmar. Experimental report of reconnaissance survey on 4–13 January 2015. Tokyo: Ramsar Center Japan.

BirdLife International (2015): Important bird areas factsheet: Indawgyi Lake Wildlife Sanctuary and surroundings. Accessible at: http://www.birdlife.org (29.06.2015).

Bryman, A. (2004): Social research methods. New York: Oxford University Press. 2 ed.

Bryman, A. (2012): Social research methods. New York: Oxford University Press. 4 ed.

Bünte, M. (2011): Burma's transition to "Disciplined Democracy". Abdication or institutionalization of military rule? GIGA Working Paper series, 177. Accessible at: https://www.giga-hamburg.de/en/system/files/publications/wp177_buente.pdf (13.08.2016).

Bünte, M. and Dosch, J. (2015): Myanmar: Political reforms and recalibration of external relations. Journal of Current Southeast Asian Affairs, 34, 2, pp. 3–19. Accessible at: http://journals.sub. uni-hamburg.de/giga/jsaa/article/viewFile/871/878 (18.08.2016).

Burma News International (2013): Deciphering Myanmar's peace process: A reference guide 2013. Accessible at: http://www.burmalibrary.org/docs14/Deciphering-Myanmar-Peace-Process-ocr-tu-red.pdf (07.08.2016).

Chambers, R. (1983): Rural development, putting the last first. London: Longman.

Chambers, R. and Conway, G. R. (1991): Sustainable rural livelihoods: practical concepts for the 21st century. IDS Discussion Paper 296. Accessible at: https://www.ids.ac.uk/files/Dp296.pdf (20.08.2014).

Chatterton, P. (2002): Be realistic: demand the impossible. Moving towards 'strong' sustainable development in an old industrial region? Regional Studies, 36 (5), pp. 552–561.

Chen, D. and Chen, H. W. (2013): Using the Köppen classification to quantify climate variation and change: An example for 1901–2010. Environmental Development, 6, pp. 69–79.

Cinner, J. (2000): Field survey questionnaire – Socioeconomic aspects of resource use and perception in Mahahual, Quintana Roo, Mexico. Kingston, Rhode Island: Department of Marine Affairs, University of Rhode Island. Accessible at: http://www.crc.uri.edu/download/CM_ Mahahualquestionaire.pdf (18.10.2013).

Cooke, P., Bokeholt, P. and Tödtling, F. (2000): The governance of innovation in Europe: regional perspectives on global competitiveness. New York: Pinter.

Data Team (2016): Myanmar in graphics: An unfinished peace (15.03.2016). Accessible at: http://www.economist.com/blogs/graphicdetail/2016/03/myanmar-graphics?fsrc=scn%2Ffb% 2Fwl%2Fbl%2Fdc%2Fanunfinishedpeace (15.03.2016).

Davies, J., Sebastian, A. C. and Chan, S. (2004): A Wetland Inventory for Myanmar. Tokyo: Ministry of the Environment, Japan.

De Lange, N. and Nipper, J. (2016): Quantitative Methoden in der Geographie. Eine Einführung. Paderborn: UTB Schöningh. (Manuscript for a textbook, will be published in 2017).

Dealtry, T. R. (1994): Dynamic SWOT analysis: developers guide. Harborne/Birmingham: Dynamic SWOT Association.

Department of Population (2015): The 2014 Myanmar Population and Housing Census: Kachin State Report (Vol. 3-A). Nay Pyi Taw: Ministry of Immigration and Population.

Effner, H., (2013): Myanmars Reformprozess. Eine Bestandsaufnahme. Accessible at: http://library.fes.de/pdf-files/iez/09668.pdf (12.08.2016).

Effner, H. and Schulz, B. (2012): Myanmar im Wandel. Leitet der Reformkurs von Präsident Thein Sein eine historische Zeitenwende ein? Accessible at: http://library.fes.de/pdf-files/iez/ 08896.pdf (12.08.2016).

Ei Ei Toe Lwin (2016): President-elect U Htin Kyaw's government has proposed slashing the number of ministries in a shake-up that would create two mega-economic ministries. Myanmar Times (17.03.2016). Accessible at: http://www.mmtimes.com/index.php/national-news/19515 -nld-to-slash-ministries-in-proposed-shake-up.html (17.03.2016).

Elliott, J. A. (2006): An introduction to sustainable development. New York: Routledge. 3 ed.

Enache, E. and Carjila, N. (2009): SWOT analysis of organic farming in Romania. Accessible at: http://papers.ssrn.com/sol3/papers.cfm?abstract_id=1517722 (30.09.2015).

Ethnic National Council of Burma (1947): Panglong Agreement. Accessible at: http://peacemaker. un.org/sites/peacemaker.un.org/files/MM_470212_Panglong%20Agreement.pdf (30.08.2016).

European Union (n.d.-a): The EC-Burma/Myanmar strategy paper (2007–2013). Accessible at: http://eeas.europa.eu/myanmar/csp/07_13_en.pdf (05.01.2015).

European Union (n.d.-b): The European Union and Myanmar/Burma – A new chapter in bilateral relations. Accessible at: http://www.euintheus.org/press-media/the-european-union-and-myanmarburma-a-new-chapter-in-bilateral-relations/ (05.01.2015).

Fei, J. C. H. and Ranis, G. (1964): Development of the labor surplus economy: Theory and policy. Homewood, Illinois: The Economic Growth Centre, Yale University.

Flick, U. (2014): An introduction to qualitative research. London: SAGE Publications. 5 ed.

Friedmann, J. (1979): On the contradictions between city and countryside. In: Folmer, H. and Oosterhaven, J. (eds.): Spatial inequalities and regional development. Boston: Martinus Nijhoff Publishing, pp. 23–46.

Gaens, B. (2013): Political change in Myanmar. Filtering the murky waters of "Disciplined Democracy". FIIA Working Paper 78. Accessible at: http://www.fiia.fi/assets/publications/ wp78.pdf (19.08.2016).

Geissmann, T., Ngwe Lwin, Saw Soe Aung, Thet Naing Aung, Zin Myo Aung, Grindley, M. and Momberg, F. (2010): Hoolock gibbon and biodiversity survey in the Indawgyi Lake Area, Kachin State and Sagaing Division, Myanmar: preliminary report. Report No. 8, Myanmar Primate Conservation Program. Biodiversity and Nature Conservation Association (BANCA), Fauna and Flora International (FFI) and People Resources and Conservation Foundation (PRCF). Yangon.

Gibbs, D. (2002): Local economic development and the environment. London: Routledge.

Hauff, M. v. (2009): Economic and social development in Burma/Myanmar. The relevance of reforms. Marburg: Metropolis-Verlag. 2 ed.

Helms, M. M. (2013): SWOT analysis framework. In: Kessler, E. H. (ed.): Encyclopedia of management theory. London: SAGE Publications.

Hla Hla Than (2006): Environmental impact of mercury contamination in some gold extraction regions. (PhD Dissertation, University of Yangon). Yangon.

Hnin Yi (2014): The political role of the military in Myanmar. RCAPS Working Paper Series "Dojo". Accessible at: http://www.apu.ac.jp/rcaps/uploads/fckeditor/publications/working Papers/RPD13003.pdf (12.08.2016).

Htoo Thant (2016): Parliament urges enquiry into last-minute rush for lucrative deals. Myanmar Times (26.02.2016). Accessible at: http://www.mmtimes.com/index.php/business/property-news/19191-parliament-urges-enquiry-into-last-minute-rush-for-lucrative-deals.html (26.02.2016).

Huang, R. L. (2012): Re-thinking Myanmar's political regime. Military rule in Myanmar and implications for reforms. SEARC Working Paper Series 136. Accessible at: http://www.cityu.edu.hk/searc/Resources/Paper/136_-_WP_-_Roger_Lee_Huang.pdf (15.08.2016).

Human Right Watch (2012): Untold miseries: wartime abuses and forced displacement in Kachin State. Accessible at: https://www.hrw.org/report/2012/03/20/untold-miseries/wartime-abuses-and-forced-displacement-burmas-kachin-state (05.06.2015).

internationalrivers (n.d): Irrawaddy Myitsone Dam. Accessible at: http://www.internationalrivers.org/campaigns/irrawaddy-myitsone-dam-0 (26.02.15).

Inter-Parliamentary Union (2016): General information about the parliamentary chamber. Accessible at: http://www.ipu.org/parline-e/reports/2388.htm (10.03.16).

Jarvis, A., Reuter, H. I., Nelson, A. and Guevara, E. (2008): Hole-filled SRTM for the globe version 4, available from the CGIAR-CSI SRTM 90m database. Accessible at: http://srtm.csi.cgiar.org (07.01.2015).

Khun Sam (2006): Sanctuary under Threat. Irrawaddy Magazine. Accessible at: http://www.irrawaddy.org/print_article.php?art_id=6396 (14.07.2011).

Kraas, F. (2014): Projektunterlagen "The Myanmar Urban Network System: 81+ cities". Yangon.

Kraas, F. and Spohner, R. (2015): Ergebnisse der Volkszählung 2014 in Myanmar. Geographische Rundschau 67 (12), pp. 44–50.

Kraas, F. and Zin Mar Than (2016): Socio-economic developments in the Indawgyi Lake Area, Kachin State, Myanmar. Journal of the Myanmar Journal of the Myanmar Academy of Arts and Science, XIV/5, pp. 281–299.

Krugman, P. (1995): Development, geography, and economic theory. Cambridge, Mass.: Massachusetts Institute of Technology Press.

Kyaw Nyunt Lwin and Khin Ma Ma Thwin (2003): Birds of Myanmar. Bangkok.

Levin, D. (2014): Searching for Burmese jade, and finding misery. (Video feature: Jade's journey marked by drugs and death). The New York Times (01.12. 2014). Accessible at: http://www.nytimes.com/2014/12/02/world/searching-for-burmese-jade-and-finding-misery.html?smid=tw-nytimes&_r=1 (30.12.2014).

Lewis, A. (1954): Economic development with unlimited supplies of labour. The Manchester School of Economic and Social Studies, 22, pp. 139–191.

Lewis, A. (1958): Unlimited labour: further notes. The Manchester School of Economic and Social Studies, 26, pp. 1–32.

Lorin, A. (2014): Myanmar's far north: The final frontier. Myanmar Time (22.12.2014). Accessible at: http://www.mmtimes.com/index.php/lifestyle/12637-myanmar-s-far-north-the-final-fronti er.html?start=1 (26.12.2014).

Maung Aung Myoe (2009): Building the Tatmadaw. Myanmar armed forces since 1948. Singapore: Institute of Southeast Asia Studies. Accessible at https://www.researchgate.net/publica tion/272092747_Building_the_Tatmadaw (12.08.2016).

McCombie, J. S. L. and Thirlwall, A. P. (1997): The dynamic Harrod Foreign Trade Multiplier and the demand-oriented approach to economic growth: An evaluation. International Review of Applied Economics, 11(1), pp. 5–26.

MCRB (Myanmar Centre for Responsible Business) and HSF (Hanns-Seidel Foundation) (2015): Community Involvement in Tourism. Nay Pyi Taw. Accessible at: http://www.myanmar-responsiblebusiness.org/news/community-involved-tourism.html (04.06.2016).

Min Zaw Oo (2014): Understanding Myanmar's peace process: ceasefire agreements. Accessible at: http://www.swisspeace.ch/fileadmin/user_upload/Media/Publications/Catalyzing_ Reflections_2_2014_online.pdf (14.07.2015).

Ministry of Agriculture and Irrigation (2002): Soil types and distribution. Accessible at: http://www.apipnm.org/swlwpnr/reports/y_ta/z_mm/mmtx221.htm#s01 (15.07.2011).

Ministry of Health (2014): Myanmar health care system. Accessible at: http://www.moh.gov.mm/ file/MYANMAR HEALTH CARE SYSTEM.pdf (18.05.16).

Morse, J. M. and Niehaus, L. (2009): Mixed method design: principles and procedures. Walnut Creek, Calif.: Left Coast Press.

Nakanishi, Y. (2007): Civil-military relations in Ne Win's Burma, 1962–1988. (PhD Dissertation, Kyoto University). Kyoto. Summary accessible at: http://www.burmalibrary.org/docs09/ Civil-Military_Relations_in_Ne_Win%92s_Burma.pdf (12.08.16).

Nakanishi, Y. (2013): Post-1988 civil-military relations in Myanmar. IDE Discussion Paper, 397, pp. 1–24. Chiba. IDE-JETRO. Accessible at: http://www.ide.go.jp/English/Publish/Download/ Dp/pdf/379.pdf (12.08.2016).

Naing Naing Latt, Kyu Kyu Thin and Seng Aung (2010): A geographical study on the socio-economic development of Indawgyi Lake environment area in Kachin State. Yangon: Dagon University Research Paper.

Neumeier, S. and Pollermann, K. (2014): Rural tourism as promoter of rural development – Prospects and limitations. Case study findings from a pilot project promoting village tourism. Accessible at: http://literatur.thuenen.de/digbib_extern/dn054583.pdf (10.10.2015).

Nyo Nyo Aung (2008): Community structure of waterbirds in Indawgyi Wetland Bird Sanctuary of Kachin State. (PhD Dissertation, University of Yangon). Yangon.

Paraskevas, A. (2013): Strengths, weaknesses, opportunities, and threats (SWOT) analysis. In: Penuel, K. B., Statler M. and Hagen, R. (eds.): Encyclopedia of crisis management. London: SAGE Publications.

Pedersen, M. B. (2016): The NLD's critical choice. Myanmar Times (17.02.2016). Accessible at: http://www.mmtimes.com/index.php/opinion/19036-the-nld-s-critical-choice.html (17.02.2016).

Pike, A., Rodriguez-Pose, A. and Tomaney, J. (2006): Local and regional development. London/New York: Routledge.

Pike, A., Rodriguez-Pose, A. and Tomaney, J. (2011): Handbook of local and regional development. London/New York: Routledge.

Ramsar (2016): Conservationists in Myanmar have a special reason to celebrate World Wetlands Day today. Accessible at: http://www.ramsar.org/news/conservationists-in-myanmar-have-a-special-reason-to-celebrate-world-wetlands-day-today (10.04.16).

Republic of the Union of Myanmar (2012): The Environmental Conservation Law (2012). Nay Pyi Taw.

Roberts, P. (2004): Wealth from waste: local and regional economic development and environment. The Geographical Journal, 170 (2), pp. 126–134.

Robinson, G. (2014): Myanmar's Transition: Economics or Politics? Which came first and why it matters. Transition Forum. Accessible at: https://lif.blob.core.windows.net/lif/docs/default-source/publications/myanmar_ned_web.pdf?sfvrsn=0 (18.08.2016).

Rodriguez, S. I, Roman, M. S. Sturhahn, S. C. and Terry, E. H. (2002): Sustainability assessment and reporting for the University of Michigan's Ann Arbor Campus. Report No. CSS02-04. Ann Arbor: University of Michigan. Accessible at: http://css.snre.umich.edu/css_doc/CSS02-04.pdf (13.01.2016).

Rubel, F. and Kottek M. (2010): Observed and projected climate shifts 1901–2100 depicted by world maps of the Köppen-Geiger climate classification. Accessible at: http://koeppen-geiger.vu-wien.ac.at/shifts.htm (30.06.15).

State Law and Order Restoration Council (1991): The Freshwater Fisheries Law. Yangon.

State Law and Order Restoration Council (1994a): The Myanmar Mines Law. Yangon.

State Law and Order Restoration Council (1994b): The Protection of Wildlife and Conservation of Natural Area Law. Yangon.

Stimson, R., Stough, R. R. and Salazar, M. (2009): Leadership and institutions in regional endogenous development. Cheltenham: Edward Elgar Publishing limited.

Stöhr, W. B. (1981): Development from below. The bottom-up and periphery-inward development paradigm. In: Stöhr, W. B and Taylor, D. R. F. (eds.): Development from above or below? The dialectics of regional planning in developing countries. Chichester/New York: John Wiley & Sons, pp. 39–72.

Stöhr, W. B. and Taylor, D. R. F. (1981): Development from above or below? Some conclusions. In: Stöhr, W. B and Taylor, D. R. F. (eds.): Development from above or below? The dialectics of regional planning in developing countries. Chichester/New York: John Wiley & Sons, pp. 453–480.

Stöhr, W. B. (1990): Introduction & On the theory and practice of local development in Europe. In Stöhr, W. B. (ed.): Global challenge and local response. Initiatives for economic regeneration in contemporary Europe. Cooperation for development series. European perspectives project, vol. 2. London: Mansell Publishing Limited, pp. 1–54.

Stöhr, W. and Tödtling, F. (1979): Spatial equity: some anti theses to current regional development doctrine. In: Folmer, H. and J. Oosterhaven (eds.): Spatial inequalities and regional development. Boston: Martinus Nijhoff Publishing, pp. 133–160.

Storper, M. (1997): The regional world: territorial development in a global economy. New York: The Guilford Press.

Sustainable tourism for development guidebook (2013): Enhancing capacities for sustainable tourism for development in developing countries. Accessible at: http://cf.cdn.unwto.org/sites/all/files/docpdf/devcoengfinal.pdf (29.12.2014).

Szirmai, A. (2005): The dynamics of socio-economic development: an introduction. Cambridge: Cambridge University Press. Accessible at: http://scholar.google.de/scholar?hl=en&q=socio-economic+development&btnG=&as_sdt=1%2C5&as_sdtp= (20.10.2013).

Taylor, R. H. (2012): Myanmar: from army rule to constitutional rule? Asian Affairs, 43 (2), 221. Accessible at: http://www.networkmyanmar.org/images/stories/PDF15/Asian-Affairs-RT.pdf (18.08.2016).

Telfer, D. J. (2002): Tourism and regional development issues. In: Sharpley, R. and Telfer, D. J. (eds): Tourism and development. Concept and issues. Clevedon: Channel View Publication, pp. 112–148.

Thant Myint-U (2009): Prepared testimony by Dr Thant Myint-U before the East Asia Sub-Committee of the Senate Foreign Relations Committee. Washington DC. Accessible at: http://www.foreign.senate.gov/imo/media/doc/Myint-UTestimony090930p.pdf (02.08.2016).

Thant Myint-U (2012): Where China meets India: Burma and the new crossroads of Asia. London: Faber and Faber.

Thant Myint-U (2016): Creating a shared identity is Myanmar's top challenge. Frontier Myanmar Magazine (07.01.2016). Accessible at: http://frontiermyanmar.net/en/interview/thant-myint-u-creating-shared-identity-myanmars-challenge (08.01.2016).

The Gender Equality Network (2012): Myanmar: women in parliament 2012. Accessible at: http://www.themimu.info/sites/themimu.info/files/documents/Ref-Doc_Myanmar-Women-in-Parliament-2012_GEN_10Oct2012.pdf (20.11.2014).

Thein, M. (2004): Economic development of Myanmar. Accessible at: https://books.google.de/books?hl=en&lr=&id=u98EBAAAQBAJ&oi=fnd&pg=PR7&dq=Thein,+M&ots=5PHYPusXFY&sig=jnTG7uKCf7_wkibsGsE5OTvxVl0#v=onepage&q=Thein%2C%20M&f=false (15.03.2016).

The Revolutionary Council of the Union of Burma (1962): The Burmese way to Socialism. Towards socialism in our own Burmese way. Rangoon. Accessible at: http://www.ibiblio.org/obl/docs/The_Burmese_Way_to_Socialism.htm (12.08.2016).

Tinker, H. (1984): Burma: the struggle for independence 1944–1948 (Vol. II). London.

Tödtling, F. (2011): Endogenous approaches to local and regional development policy. In: Pike, A. Rodriguez-Pose, A. and John, T. (eds.), Handbook of local and regional development. New York: Routledge, pp. 333–343.

Tyn Myint-U (2010): Developing Myanmar: toward a knowledge-based economy. Deer Park, NY: Linus Publications.

UNESCO (2016): Ecological sciences for sustainable development. Man and biosphere programme. Accessible at: http://www.unesco.org/new/en/natural-sciences/environment/ecological-sciences/man-and-biosphere-programme/ (10.04.16).

UNESCO World Heritage Center (2014): Indawgyi Lake Wildlife Sanctuary. Accessible at: http://whc.unesco.org/en/tentativelists/5872/ (10.04.16).

UNICEF (2013): Snapshot of social sector public budget allocations and spending in Myanmar. Accessible at: https://www.unicef.org/myanmar/Final_Budget_Allocations_and_Spending_in_Myanmar.pdf (18.05.2016).

Union of Myanmar. (2008): Constitution of the Republic of the Union of Myanmar (2008). Nay Pyi Taw: Ministry of Information.

Vazquez-Barquero, A. (1999): Inward investment and endogenous development. The convergence of the strategies of large firms and territories? Entrepreneurship & Regional Development, 11(1), pp. 79–93.

Vazquez-Barquero, A. (2003): Endogenous Development: Networking, innovation, institutions and cities. London/New York: Routledge.

Wa Lone (2016): Guns fall silent in Kyaukme, but tension high on the front lines. Myanmar Times (22.02.2016). Accessible at: http://www.mmtimes.com/index.php/national-news/19097-guns-fall-silent-in-kyaukme-but-tension-high-on-the-front-lines.html (22.02.2016).

Wilson, T. (2015): Is Myanmar's nationwide ceasefire agreement good enough. East Asia Forum (21.10.2015). Accessible at: http://www.eastasiaforum.org/2015/10/21/is-myanmars-nation wide-ceasefire-agreement-good-enough/ (09.04.2016).

World Commission on Environment and Development (1987): Our common future. Oxford: Oxford University Press.

World Development Indicators (2013): Expenditure on education, public (% of GDP) (%). Accessible at: http://hdr.undp.org/en/content/expenditure-education-public-gdp (18.05.16).

Ye Mon and Verbruggen, Y. (2016): At one-year anniversary, stalled trails a democracy litmus test. Myanmar Times (10.03.2016). Accessible at: http://www.mmtimes.com/index.php/national-news/19399-at-one-year-anniversary-stalled-student-trials-a-democracy-litmus-test.html (17.03.2016).

Zin Mar Than (2011): Socio-economic analysis of the Indawgyi Lake Area, Mohnyin Township, Myanmar. Master thesis, Cologne University. Accessible at: http://library.fes.de/pdf-files/stufo/cd-0768/thesis.pdf (10.01.2014).

APPENDICES

APPENDIX 1: HOUSEHOLD QUESTIONNAIRE

University of Cologne, Germany　　　　　　　Institute of Geography

Household Questionnaire
Socio-Economic Development Potentials in Indawgyi Lake Area, Kachin State, Myanmar
Zin Mar Than
PhD candidate, DAAD (German Academic Exchange Service) scholar

Aim of the study: to investigate the current socio economic situation in Indawgyi Lake Area and to find out and discuss what kind of potential for the future development does the area have.

Respondent name: ....

Ethnic group of respondent: ....

Village Name: ....　Village Tract: ....

Date: ....　Tel: ....

1.　How many members live in the household currently?

2.　Please, give me the following information about all the members, currently living in your household

no	HH member (relationship)	sex	age	occupation	lace of birth

3.　Has this household migrated during the last 10 years?

1 from where .... 2 when .... 3 *reason .... (see no. 4)

4.   If some household members have migrated during the last 10 years, please give me the following information for these former members.

HH member	to where	year left	*reason

*reason is main reason
1. work   2. education   3. illness   4. marriage
5. to escape insecurity   77. don't know   88. no answer   99. other...................

Health

5.   What is the distance to the hospital/ health centre?

☐ 77 don't know   ☐ 88 no answer

6.   Do you usually sleep under a mosquito net?
☐ 1 yes   ☐ 2 no   ☐ 77 don't know ☐ 88 no answer

7.   Did any household member fall sick or got injured during the last 60 days?
☐ 1 yes   ☐ 2 no   ☐ 77 don't know ☐ 88 no answer

8.   Health treatment, if yes

HH member	how many days?	treatment place?	if no external treatment place was used, main reason	type of ownership of treatment place	costs for drugs	in case of costs, why? no cost, why?
	77 don't know 88 no answer	1 none/home 2 hospital 3 dispensary 4 health centre 5 drug shop 6 traditional doctor 77 don't know 88 no answer 99 other...........	1 illness mild 2 facilities are too far 3 costs 4 staff not available 5 staff attitude not good 6 drugs not available 77 don't know 88 no answer 99 other...........	1 government 2 religious/ NGO 3 private (for profit) 77 don't know 88 no answer 99 other..........	1 yes 2 no 77 don't know 88 no answer	if yes, 11 some drugs had to be purchased 12 all drugs had to be purchased if no 21 no drugs required 22 obtained drugs free of charge

9. Education

HH-member	Schooling status	if never attended main reason	left school main reason	current attending	level attained	read & write	attend literacy program
	1 never attended 2 left school 3 currently attending 77 don't know 88 no answer	1 too young 2 sick 3 disable 4 need to work 5 costs 6 school too far 7 orphaned 8 indifferent 9 insecurity 77 don't know 88 no answer 99 other..........	1 completed a desired level 2 need to work 3 costs 4 transport 5 quality of school 6 orphaned 7 sickness 8 calamity in family 9 got married 10 take care of relatives/house work 11 indifferent 12 insecurity 77 don't know 88 no answer 99 other..........	1 nursery 2 primary 3 secondary 4 high school 5 university 6 distance university 77 don't know 88 no answer 99 other.......	1 monastic school 2 primary 3 secondary 4 high school 5 university 77 don't know 88 no answer 99 other..........	1 neither able to read nor write 2 able to read only 3 able to read & write 77 don't know 88 no answer 99 other.......	1 yes 2 no 77 don't know 88 no answer

10. What is the distance to school?

☐ 1 primary   ☐ 2 secondary   ☐ 3 high school

☐ 4 university   ☐ 77 don't know   ☐ 88 no answer

11. Housing Condition

house material			type of kitchen	type of bathroom	type of toilet
roof	wall	floor			
1 iron sheet 2 bamboo 3 thatch 77 don't know 88 no answer 99 other.................	1 wood 2 bamboo 3 brick 77 don't know 88 no answer 99 other.............	1 wood 2 bamboo 3 brick 77 don't know 88 no answer 99 other.............	1 inside 2 outside (makeshift) 3 outside (built) 4 none 77 don't know 88 no answer 99 other.................	1 inside 2 outside (makeshift) 3 outside (built) 4 none 77 don't know 88 no answer 99 other.............	1 covered pit latrine 2 uncovered pit latrine 77 don't know 88 no answer 99 other.................

12. Living Condition

main fuel source		main water source		solid waste disposal main practice
lighting	cooking	drinking	HH water	
1 electricity 2 candle 3 lamp 4 generator 5 solar 77 don't know 88 no answer 99 other..................	1 firewood 2 charcoal 3 electricity 77 don't know 88 no answer 99 other....................	1 lake 2 tube well 3 stream 4 rain water 77 don't know 88 no answer 99 other..................	1 lake 2 tube well 3 .stream 4 rain water 77 don't know 88 no answer 99 other..................	1 burning 2 pit 3 heap 4 garden/compost 77 don't know 88 no answer 99 other..................

13.   What is the household's main source of power supply?
☐1 public source   ☐2 private source   ☐77 don't know
☐88 no answer   ☐99 other.....................

14.   What is the distance to?

☐ drinking water source   ☐ household water source
☐77 don't know   ☐88 no answer

15.   Does this household own any of the following transportation devices?

device	unit	since when
motor bike		
bike		
boat		
motor boat		
bullock cart		
motor vehicle		
other		

16.   What is the household's main source of information?
☐1 print media   ☐2 electronic media (e.g. TV, Radio)   ☐3 word of mouth
☐4 post mail   ☐77 don't know ☐88 no answer   ☐99 other.................

17.   Does this household own any of the following communication equipment?

equipment	unit	since when
mobile phone		
line phone		

18.   Does this household own any of the following entertainment equipment?

equipment	unit	since when
TV		
VCD/DVD		
cassette recorder		
radio		
computer		

Household Expenditure

19.   Expenditure for short terms – food, drink and tobacco (during last week)

description	purchase	home production	market price
rice			
noodle/ vermicelli			
meat			
fish			
others (egg,...)			
fresh vegetable (cabbage...)			
ingredient			
fruit			
tea/coffee			
smoking			
alcoholic			
other			

20.   Expenditure for mid-term non-food (during last 30 days)

description	purchase	home production	market price
power			
fuel (firewood, charcoal...)			
bathing material (soap...)			
print media (newspaper...)			
benzine			
bus/boat fare			
service for mobile phone			
health			
social contribution			
other			

21.   Expenditure for long-term non-food (during last 365 days)

description	purchase value	home produce value
clothes (men, women, children)		
bedding material		
household equipment		
travelling cost		
repair costs		
jewel		
cost for education		
others expenses		

Economic Structure

22.   What are the three main economic activities of your household?

☐ 1 farming   ☐ 2 fishing   ☐ 3 retail shop
☐ 4 self-employed   ☐ 5 casual/agriculture labour ☐ 6 staff (government/private)
☐ 7 Livestock breeding ☐ 8 trade   ☐ 77 don't know
☐ 88 no answer   ☐ 99 other..............................

23.   If the activity is farming, please give me the following information.

23.1   Since when is farming the economic activity?

23.2   How many acres do you have?

23.3   How many of these acres are own farming land?

23.4   Farming products and generated income

what kind of crops	how many harvesting times/y	average yield/y	selling price/unit	selling practice

23.5 Investment for farming per year

seed cost/y	fertilizer/ organic costs/y	pesticide costs/y	irrigation costs/y	machine costs/y	machine rental costs/y	labour costs/y	interest for loan/y

24.   If the activity is fishing, please give me the following information.

24.1   Since when is fishing the economic activity?

24.2   What kind of fishing gear do you use mainly?

24.3   How is the fishing practice?

24.4   Fishing products and generated income

kind of fish	caught fish/day	selling price	selling practice

24.5   Investment for fishing per year

boat	fuel cost/y	fishing gear	interest for loan/y

25.   If the activity is retail shop, please give me the following information.

25.1   Since when is this business the economic activity? ☐

25.2   What kind of goods do you trade mainly? ☐

25.3   How is the trading practice? ☐

25.4   Income per year from retail shop

average sale volume per day	profit %	ave. profit/day

25.5   Investment per year for retail shop

average investment	interest for loan/y	other costs/y

26.   If the activity is self-employment (gold miner, carpenter...), please give me the following information.

26.1   Since when is self-employment the economic activity? ☐

26.2   Income per year (average/y)

working day/month x 12	average income/month x 12	average income

26.3   Investment per year

average investment	interest for loan/y	other costs/y

27.   If the activity is casual labour/agriculture labour, please give me the following information.

27.1   Since when is casual labour/agriculture labour the economic activity? ☐

27.2   Income per year

working day/month x 12	daily wages	average income	interest for loan/y

28.   If the activity is government / private staff, please give me the following information.

28.1   Since when is government / private staff the economic activity? ☐

28.2   Income per year

salary/month x 12	average income	interest for loan/y

29.   Livestock breeding

kind of livestock	how many of this livestock are sold/y	selling price	average income/y	investment for livestock

30.   If the activity is trade, please give me the following information

30.1   Since when is trade the economic activity?

30.2   What kind of goods do you trade mainly?

30.3   How is the trading practice?

30.4   Income per year (average/y)?

30.5   Investment per year

average investment	interest for loan/y	other costs/y

30.6   Has the household other income than mentioned before?

31.   Evaluation of current socio-economic situation

dimension	good	tending to good	tending to bad	bad
income generating				
job opportunity				
market accessibility				
transportation facility				
communication facility				
information facility				
education facility				
health facility				
electricity system				
waste management				
farm water availability				
water quality				
impact of conservation area on local economy				
eco-tourism				

32.   Future

32.1   Do you have plans for coming five years to change the economic activities/ base of your household?

☐ 1 no. a change is not needed

☐ 2 yes, we have plans to change something, but we will keep the present economic base

☐ 3 yes, we have plan to change and will give up the present economic base almost completely and will establish a new economic base

☐ 77 don't know   ☐ 88 no answer

If the answer is yes, no. 2 changing something:

32.2   What kind of changes do you have in mind mainly? Please give only a maximum two answers.

(1) ....

(2) ....

If the answer is yes, no. 3 giving up the present economic base:

32.3   What is the reason for giving up?
☐ 1 the income generation based on the present activity is not good enough
☐ 2 the work in the present field is too hard
☐ 3 the young people in the household do have other ideas and interests
☐ 77 don't know   ☐ 88 no answer   ☐ 99 other................................

32.4   What kind of new economic base the household will establish?
Pls. name the main activities (max.3)
☐ 1 agriculture/farming ☐ 2 fishing   ☐ 3 trade
☐ 4 shop keeping   ☐ 5 service (tourist sector) ☐ 6 service (general)
☐ 77 don't know   ☐ 88 no answer   ☐ 9 other.................

32.5   What do you think, how will your household situation develop in the coming five years?

dimension	very positive	positive	negative	very negative
income generating				
expenses				
housing condition				
hygienic condition				
impact of conservation area on HH economy				

32.6   What do you think, how will the situation in your village change in the coming five years?

dimension	very positive	positive	negative	very negative
income generating				
job opportunity				
market accessibility				
transportation facility				
communication facility				
information facility				
education facility				
health facility				
electricity system				
waste management				
impact of conservation area on local economy				
eco-tourism				

32.7   Are you interested in eco-tourism sector?
☐ yes   ☐ no

32.8 If yes, what do you think, how can you be involved into eco-tourism?

(1) ....

(2) ....

32.9 What do you think are the three factors/items, which are the most important ones having to be improved in the next years in your village?

(1) ....

(2) ....

(3) ....

APPENDIX 2: KEYWORDS FOR THE EXPERT INTERVIEWS

The interviews with the experts are focused on three main fields: infrastructure, socio-demographic matters and economy. These fields are subdivided into subareas. Dependent on the experts and their specialisation not all topics are covered always, but only on the ones in which the expert are specialized.

All interviews are based on the idea of the SWOT concept, which implies that in every interview the experts were asked for
–   today strengths,
–   today weaknesses,
–   future opportunities,
–   future threats
within the fields and subareas they are talking about. Also every interviewee is asked for possible solutions and the question: which are the three most important things, which are urgently needed to improve the Indawgyi Lake Area.

The keywords for the interviews are listed below.

Field: infrastructure
–   Subarea: Road transportation/water transportation
    •   road situation for the whole year
    •   mode of transportation in the area
    •   traffic volume
    •   connection with the region
    •   comparing past and present situation of roads, transportation lines
    •   investment in road construction
    •   future plan for road transportation development
–   Subarea: electricity supply
    •   situation
    •   sources
    •   investment in electricity
    •   future plans and requirements
–   Subarea: water supply
    •   type of water supply system (lake, tube well, rain water)
    •   quality of water
    •   seasonal differences
    •   practice of rainwater harvesting
    •   needs to improve for water sources for future
–   Subarea: waste disposal management
    •   general situation of waste
    •   disposal system: where? how?
    •   flooding events: where? when?
    •   future plans and requirements

Field: socio-demographic matters
- Subarea: demography, general
  - total population, population distribution, density
  - ethnic groups
  - future trends
- Subarea: migration
  - migrants: amount, place of origin
  - driving factors of migration
  - places for settling, why there?
  - migrant settlements: systematic? informal?
- Subarea: education, general
  - situation of schools, number of primary/middle/high schools
  - situation of buildings
  - facilities: library, playground etc.
  - teachers' situation: salary level, qualification level
  - student/teacher ratio
  - curriculum about ethnic language
  - entry fees, private tuition situation
- Subarea: higher education (university and colleges)
  - situation at the university, major specialization
  - situation of buildings
  - number and origin of students
  - qualification of students and teachers
  - needs of the institution
  - future development
- Subarea: health
  - situation of health care: public health centre and clinic
  - quality and facilities (building condition, equipment) of public health centre and clinic
  - amount of manpower
  - kind of diseases, causes of diseases, patients are from where

Field: economy
- Subarea: agriculture
  - main products
  - number of harvests per year
  - yields: their quantities and their qualities, future enhancement
  - usage of fertilizer, pesticide, and irrigation system, financial support (loan) ? existence of education programs for that?
  - agricultural technology: kind of technology used? changes from traditional methods to mechanized farming (when) ?
  - introduction of new products
  - market accessibility
  - labour force

- Subarea: fishery
  - fishing regulation (permits of fishing, kind of fishing gear)
  - situation of fishermen per village (in particular migrant fishermen)
  - kind of fish species and rebreeding program
- Subarea: industry
  - kind of agro-based (rice mill, food processing) and non-agro based (ice mill and gold mining) industry
  - development of industrial sector
  - opportunities to develop the industrial sector
- Subarea: handicraft production
  - handicrafts: kind, type, products, skill
- Subarea: services
  - kind, types (restaurant and casual labour)
  - connection to Hpakant
  - market demand and market situation
  - measures to improve the situation
- Subarea: tourism and recreation
  - situation of domestic and international tourism
  - tourists: when?, from where?, why?
  - duration of stay
  - condition of tourism: number of beds and rooms, quality of accommodation
  - local guides: number, qualification
  - safety for visitors
  - interest of visitors, reason of interest
  - attractions: tracking trail, bird watching, other flora and fauna.... where and how do people enjoy?
  - future development directions in tourism
  - What does make people particularly happy, when do they think of Indawgyi Lake Area?

Field: conservation
  - man power and facilities
  - regulation for conservation
  - zoning system (in the lake and on the land)
  - controlling system (problems, intensity)
  - education program
  - funding
  - collaboration with non-governmental organization

APPENDIX 3: LIST OF INTERVIEWEES

Interviews from January to May 2014

Code	Institution/Activity	Position	Interview location
IDGY-01	Hospital, Loneton	Medical doctor	Loneton
IDGY-02	Indaw Mahar Guest House	Clerk	Loneton
IDGY-03	Fishery Department	Assistance officer	Loneton
IDGY-04	Private Clinic	Medical doctor	Nammilaung
IDGY-05	Indawgyi Wildlife Sanctuary	Ranger	Loneton
IDGY-06	Regional office	District officer	Mohnyin
IDGY-07	Administration Department	Township officer	Mohnyin
IDGY-08	Indomyanmar Conservation (NGO)	Head of IMC	Yangon
IDGY-09	Indomyanmar Conservation (NGO)	Project coordinator	Yangon
IDGY-10	Fauna &Flora International (NGO)	Conservation coordinator	Yangon
IDGY-11	Deutsche Welthungerhilfe(NGO)	Country director	Yangon
IDGY-12	Middle school	Head of school	Lonsent
IDGY-13	Farming	Village elder & farmer	Lonsent
IDGY-14	Woman affair	Member	Lonsent
IDGY-15	Farming	Village elder & farmer	Nammokkam
IDGY-16	Middle school	Head of school	Nammokkam
IDGY-17	Health centre	Health assistant	Nammokkam
IDGY-18	Village Admin	Village head	Nammokkam
IDGY-19	Health centre	Midwife	Hepa
IDGY-20	Village Admin	Village elder & head	Hepa
IDGY-21	Village Admin	Village head	Nyaungbin
IDGy-22	Health centre	Midwife	Nyaungbin
IDGY-23	Village Admin	Village head	Nammilaung
IDGY-24	Village Admin	Village head	Nanpade
IDGY-25	Indawgyi Wildlife Sanctuary	Warden	Mohnyin
IDGY-26	Mohnyin Degree College	Principle	Mohnyin
IDGY-27	Youth Association	Member	Mamomkai
IDGY-28	Farming	Village elder & farmer	Mamomkai
IDGY-29	Health centre	Midwife	Mamomkai
IDGY-30	Village Admin	Village head	Mamomkai
IDGY-31	Village Admin	Village head	Loneton
IDGY-32	Education	Retiree & village elder	Main Naung
IDGY-33	High School (branch)	Head of school	Main Naung
IDGY-34	Village Admin	Village head	Main Naung
IDGY-35	Health centre	Midwife	Main Naung
IDGY-36	Village Admin	Village head	Shweletpan
IDGY-37	Cologne University, Germany	Professor	Brühl, Germany

Interviews in December 2015

Code	Institution/Activity	Position	Interview location
IDGY-38	Kachin Independence Org. (KIO)	Spoke person	Myitkyina
IDGY-39	Agriculture Dept. of Kachin State	Deputy director	Myitkyina
IDGY-40	Myitkyina Hospital	Head of hospital	Myitkyina
IDGY-41	Basic Education for Kachin State	Director	Myitkyina
IDGY-42	Wildlife Sanctuary	Warden & ranger	Myitkyina
IDGY-43	Mohnyin Fishery Department	District officer	Mohnyin
IDGY-44	Administration Department	Township officer	Mohnyin
IDGY-45	Mohnyin Degree College	Principle	Mohnyin
IDGY-46	Friends of Wildlife (NGO)	Educationist	Loneton
IDGY-47	Indawgyi Wildlife Sanctuary	Warden	Loneton
IDGY-48	Loneton Fishery Office	Assistant fishery officer	Loneton
IDGY-49	Mohnyin Degree College	Professor	Loneton
IDGY-50	Ministry of Forestry	Director	Nay Pyi Taw
IDGY-51	Ministry of Hotel and Tourism	Officer	Myitkyina
IDGY-52	Myitkyina Agriculture Office	Officer	Myitkyina
IDGY-53	Land and Land Record Department	Assistant director	Myitkyina
IDGY-54	Kachin State	Peace negotiator	Myitkyina

Alexander Follmann

Governing Riverscapes

Urban Environmental Change along the
River Yamuna in Delhi, India

MEGACITIES AND GLOBAL CHANGE / MEGASTÄDTE UND
GLOBALER WANDEL — VOL. 20

Existing research analyzes riverfront developments largely from a
city-centric point of view, assuming a clear boundary between the
river and the city. The research presented in this book shows that
the complexity of urban environmental transformation
along rivers in the megacities of the Global South requires a
change of perspective, going beyond such a dichotomous view.
By linking a discourse analytical approach with concepts from
governance research and urban political ecology, this study introduces the theoretical framework of riverscapes as socio-ecological
hybrids for a comprehensive analysis. The concept is applied to
the river Yamuna.
Delhi's riverscapes have recently seen large-scale slum demoli-
tions and the development of urban mega-projects. These dynamic land-use changes are deeply connected to changing discursive
framings of the role and function of Delhi's riverscapes in the
remaking of the megacity. The study shows how dominant dis-
courses and their associated narratives have remained persistent
over long periods of time and the influence they continue to have
on urban environmental change and governance.

2016
396 pages with 23 col. and
26 b/w photos, 7 col. and
46 b/w illustrations, 13 tables
and 6 col. maps
978-3-515-11430-1 SOFTCOVER
978-3-515-11435-6 E-BOOK

Franz Steiner
Verlag

Please order here:
www.steiner-verlag.de

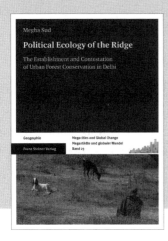

Megha Sud

Political Ecology of the Ridge

The Establishment and Contestation of
Urban Forest Conservation in Delhi

MEGACITIES AND GLOBAL CHANGE / MEGASTÄDTE UND
GLOBALER WANDEL – VOL. 23

The remains of the time-worn Aravalli Mountains that extend
through Delhi are known as the Ridge. The range is also the site of
Delhi's once continuous but now scattered forests, four segments
of which are legally protected as wildlife conservation areas
today. How did a biodiversity park and a wildlife sanctuary come
to be established in a large and rapidly growing megacity with
severe competition for land resources? What does this mean for
various citizens who use the forests for different purposes?
Megha Sud provides an insight into the answers to these questions by examining the negotiations that led to the establishment
of conservation as the dominant discourse in the Ridge, highlight-
ing the role of various actors and the unequal power relations
embedded in these deliberations and their outcomes. In this study
she brings together concepts from political ecology and urban
political ecology, using literature on local politics and development to ground the argument in the context. She argues that the
forested spaces of Delhi must be understood as implicated in the
socio-political structures of the city if urban conservation policies
are to be sustainable both socially and environmentally.

2017
267 pages with 10 b/w-photos,
3 b/w-illustrations, 6 maps and
3 tables
978-3-515-11714-2 SOFTCOVER
978-3-515-11715-9 E-BOOK

Franz Steiner
Verlag

Please order here:
www.steiner-verlag.de

Frauke Kraas / Regine Spohner / Aye Aye Myint

Socio-Economic Atlas of Myanmar

In collaboration with Aung Kyaw, Hlaing Maw Oo, Htun Ko, Khin Khin Han, Khin Khin Soe, Myint Naing, Nay Win Oo, Nilar Aung, Saw Yu May, Than Than Thwe, Win Maung, Zin Mar Than, Zin Nwe Myint

The Socio-Economic Atlas of Myanmar focuses on the analysis and evaluation of regional differences in geographical conditions, natural resources, infrastructure and in particular the socio-economic development in the states and regions of the country in the current transformation process of Myanmar. The atlas is based on international literature, statistical data, qualitative research and spatial information within a Geographic Information System on Myanmar. The spatial analyses aim to increase the state of knowledge about Myanmar both within the country and abroad, and to support decision-making on spatial development policy.

CONTENTS
Concept and data of the atlas | Administrative and spatial organisation | Population, settlements and urbanisation | Infrastructure | Economic development | Social development: household infrastructure, education and health | References

2017
Size 21,0 x 29,7 cm
188 pages with 52 colour photos, 13 colour tables, 54 colour illustrations
978-3-515-11623-7 HARDCOVER
978-3-515-11625-1 E-BOOK
OPEN ACCESS, VIA
ELIBRARY.STEINER-VERLAG.DE

Franz Steiner Verlag

Please order here:
www.steiner-verlag.de